SpringerBriefs in Agriculture

More information about this series at http://www.springer.com/series/10183

Waseem Mushtaq
Mohammad Badruzzaman Siddiqui
Khalid Rehman Hakeem

Allelopathy

Potential for Green Agriculture

 Springer

Waseem Mushtaq
Department of Botany
Aligarh Muslim University
Aligarh, Uttar Pradesh, India

Mohammad Badruzzaman Siddiqui
Department of Botany
Aligarh Muslim University
Aligarh, India

Khalid Rehman Hakeem
Department of Biological Sciences
King Abdulaziz University
Jeddah, Saudi Arabia

ISSN 2211-808X ISSN 2211-8098 (electronic)
SpringerBriefs in Agriculture
ISBN 978-3-030-40806-0 ISBN 978-3-030-40807-7 (eBook)
https://doi.org/10.1007/978-3-030-40807-7

This Springer imprint is published by the registered company Springer Nature Switzerland AG
The registered company address is: Gewerbestrasse 11, 6330 Cham, Switzerland

Foreword

Allelopathic studies may be defined in various aspects: weed against weed/crop and vice versa. The book focuses on the ways to utilize the allelopathic potential of weeds or crops for controlling weeds in the agroecosystems. Vigorous use of herbicides is poisoning our environment at an alarming rate; allelopathy can be employed as a useful alternative to control weeds naturally under field conditions. The book contains chapters on the history of allelopathy: allelopathic potential of several important crops (rice, wheat, sorghum, maize, mustard, sunflower) and weeds (members of Solanaceae, Convolvulaceae, Asteraceae, Verbenaceae). Moreover, it highlights how the allelopathic potential of these weeds and crops can be employed effectively to suppress weeds under field conditions. The book also discusses topics on the role of allelochemicals in agroecosystems, impact on local flora, biotic stress induced by allelochemicals, mechanism of action of allelochemicals, and future prospective of allelopathy.

The book is a repository of advanced research in allelopathy. It shall act as a useful reference for the latest advances in research in the concerned area. The content of the book is diverse and shall attract the attention of students, researchers, and scientists world over.

Department of Biological Sciences   Khalid Rehman Hakeem
King Abdulaziz University
Jeddah, Saudi Arabia

Preface

This monograph is the first of its kind that provides an in-depth insight into allelopathic science. No general monograph on the subject of allelopathy has been published previously that discusses the mechanism behind this branch of study. The wide recognition by agro-ecologists of allelopathy as a vital 'plant interactions' phenomenon has followed only during the last decade. Therefore, there seems to be a requirement for a broad reference source in this arena, both for related researchers in the field and as an outline for allelopathic aspirants. Most substantial contributions in the subject, accessible at the time of inscription, have been conferred. The basic aim is to discuss the comprehensive ecological roles where allelopathy can be employed for green agriculture. All levels of biochemical interactions between weeds and crops discussed are intricately entwined in ecological phenomena. Most of my own research during my Ph.D. has been reported here, for which I am grateful. I deeply acknowledge the enthusiastic motivation of my parents, collaborators, and friends without whose help this monograph would have only been a dream.

Aligarh, India   Waseem Mushtaq

Contents

1	**Introduction**		1
	References		3
2	**History of Allelopathy**		5
	2.1	Methodological Approaches Used in Allelopathic Studies	6
	2.2	Reports on Allelopathic Plants from 2010	7
	References		19
3	**Allelopathy Potential of Important Crops**		25
	3.1	Rice	25
	3.2	Wheat	27
	3.3	Sorghum	29
	3.4	Maize	31
	References		32
4	**Allelopathy Potential of Weeds Belonging to the Family**		37
	4.1	Asteraceae	37
	4.2	Convolvulaceae	38
	4.3	Solanaceae	39
	4.4	Verbenaceae	39
	References		41
5	**Role of Allelochemicals in Agroecosystems**		45
	References		49
6	**Allelopathic Control of Native Weeds**		53
	6.1	Allelopathic Crops	54
	6.2	Allelopathic Compounds	54
	6.3	Allelopathic Weed Control	55
	6.4	Approaches for Allelopathic Weed Management	55
		6.4.1 Intercropping	55
		6.4.2 Cover Crops	56
		6.4.3 Residue Incorporation	56
	References		57

7   Mechanism of Action of Allelochemicals.   61
      References.   64

8   Future Prospective   67
      References.   69

Chapter 1
Introduction

Weeds persistently contend with crops causing a substantial loss in their yield (Mushtaq & Siddiqui, 2018). Losses caused by weeds are far above the losses from any category of agronomic pests such as insects, diseases, nematodes, rodents, etc., (Abouziena & Haggag, 2016). An average reduction of 34% is caused by weeds in crop production (Oerke, 2006). Some commercial crops that suffer reductions in their harvest due to weeds are as follows: wheat 23%, potatoes 30%, cotton 36%, rice 37%, soybeans 37%, and maize 40% (Oerke, 2006). Weeds acquire a serious share of applied fertilizers and reduce their accessibility to crops (Bajwa, 2014; Guglielmini, Verdu, & Satorre, 2017). They restrain the crop plants to the available light, moisture, and space as well (Guglielmini et al., 2017). Moreover, they decline the standard of crops, obstruct water channels, disturb human health, cause fire threats, and appear unpleasant in recreation areas like gardens, pools, parks, pavements, and pathways (Singh, Batish, & Kohli, 2003). Therefore, weeds are recognized as severe plant pests ever since the ancestral days (Zimdahl, 2013). They have constantly engaged with the agricultural practices which made it necessary to adopt certain measures to look for their control (Zimdahl, 2013).

Since the commencement of agriculture, hand weeding, motorized weeding, and herbicide applications have been the most prominent conventional weed control approaches (Chauvel, Guillemin, Gasquez, & Gauvrit, 2012; Jabran, Mahajan, Sardana, & Chauhan, 2015; Young, Pierce, & Nowak, 2014). Such weed control measures have assisted to keep weed saturation low thereby improving crop production throughout the globe. In spite of the significant contribution of the conventional weed control methods, some problems are also concomitant to them, which makes it very imperative to develop mature assortment in the present weed control measures. Further, the costs incurred upon weed eradication are also enormous; weeds cause about 12% loss in the yield in the USA and it costs around US$ 35 billion for their control (Pimentel et al., 2001). Heady and Child (1999) observed that in the U.S. Department of Agriculture during the 1960s, control and management of land invasion by weeds cost US$ 2.5 billion dollars yearly on western United States

© The Author(s), under exclusive license to Springer Nature Switzerland AG 2020
W. Mushtaq et al., *Allelopathy*, SpringerBriefs in Agriculture,
https://doi.org/10.1007/978-3-030-40807-7_1

rangelands, followed by US$ 340 million dollars in 1989 in seventeen western united. Apart from the direct losses, roughly $4 billion is spent each year to control weeds through weedicides/pesticides (Inderjit, Seastedt, Callaway, Pollock, & Kaur, 2008) and the costs maybe even more in developing and under-developed nations. As far as yield loss due to weeds in India is concerned, reduction in crop yield was estimated at 31.5% by weeds (Anonymous, 2007). In another study, it was reported that loss in agricultural productivity due to weeds amounts to INR 1050 billion per annum (Varshney & Babu, 2008).

Considering these elements of weeds and their harms, it ends up necessary to control them. In this manner, efforts are being made to discover elective low-input methodologies for weed management, despite many weed control rehearses are accessible. The increasing challenges attached to weed management like herbicide-resistant weeds (Harker, 2013) and chemical leftovers of herbicides widen the scope for organic crops; and growing concern over human health issues and environmental (Vyvyan, 2002) needs an in-depth farming structure that is less chemical-dependent and supports natural compounds. Allelopathy has profound implications for weed suppression and is encompassed among the significant prolific weed control measures (Jabran & Farooq, 2013; Zeng, 2014). Allelopathy is an ecological phenomenon by which a plant affects the growth behaviour, physiology, and development of another plant living in its proximity through the release of some chemical/s (Farooq, Jabran, Cheema, Wahid, & Siddique, 2011; Hussain, Gonzalez, & Reigosa, 2011; Pan, Li, Yan, Guo, & Qin, 2015). With this perspective, Khanh, Chung, Xuan, and Tawata (2005) presented allelopathic plants for sustainable agricultural production, and their products depicting herbicidal, antifungal, insecticidal, and pesticidal properties have received much attention (Javed, Javaid, & Shoaib, 2014; Ladhari, Omezzine, Dellagreca, Zarrelli, & Haouala, 2013). Allelopathic plants releasing allelochemicals into the environment are biodegradable (Duke et al., 2002), thus cause less pollution, safeguard agronomic produces (Sodaeizadeh, Rafieiolhossaini, & Van Damme, 2010), along with alleviating social health problems (Khanh, 2007) and therefore are a better option to replace chemical herbicides for weed management. The allelopathic nature of about 240 weed species is reported that interfere with the growth and production of crops (Qasem & Foy, 2001). The control of weeds through allelopathy may be effective as a solitary technique in certain cropping frameworks, for example, organic farming. Moreover, it can be joined with other farming methods to accomplish integrated weed management. Under the practice of weed control through allelopathy, the allelopathic competence of crops is utilized as such that the allelochemicals present in these crops decrease weed infestation. The plants either living or their dead counterparts express allelopathy through allelochemical(s) exudation. Weed control through allelopathy can be executed by developing allelopathic plants in close proximity to weeds which endorse the generation of these chemical(s) (Tesio & Ferrero, 2010) or by setting the allelopathic materials acquired from dead plants near weeds. The decaying plant products discharge allelochemicals which are consumed by the objective weeds. The most imperative case for such cases incorporates the utilization of allelopathic plant remains for controlling weeds (Tabaglio, Marocco, & Schulz, 2013). Allelopathic weed control can likewise be applied by developing allelopathic plants in a field for a specific time-frame, all together for their underlying roots to radiate allelochemicals. The most vital case of such allelopathic weed control is crop rotation (Farooq et al., 2011). Another allelopathic approach to filter weeds incorporates acquiring allelochemicals in a fluid system by dunking the allelopathic chaff in water for a specific timeframe. A few studies have pushed utilizing along these lines of weed control either alone or in integration with different weed control measures (Jabran et al. 2013, 2015). In this manner, for the management of agrarian weeds, it is beneficial to investigate the solid allelopathic potential of the plant.

References

Abouziena, H. F., & Haggag, W. M. (2016). Weed control in clean agriculture: A review. *Planta Daninha, 34*(2), 377–392.

Anonymous. (2007). *Vision 2025. NRCWS perspective plan*. New Delhi, India: Indian Council of Agricultural Research (ICAR).

Bajwa, A. A. (2014). Sustainable weed management in conservation agriculture. *Crop Protection, 65*, 105–113.

Chauvel, B., Guillemin, J. P., Gasquez, J., & Gauvrit, C. (2012). History of chemical weeding from 1944 to 2011 in France: Changes and evolution of herbicide molecules. *Crop Protection, 42*, 320–326.

Duke, S. O., Dayan, F. E., Rimando, A. M., Schrader, K. K., Aliotta, G., Oliva, A., & Romagni, J. G. (2002). Chemicals from nature for weed management. *Weed Science, 50*(2), 138–151.

Farooq, M., Jabran, K., Cheema, Z. A., Wahid, A., & Siddique, K. H. (2011). The role of allelopathy in agricultural pest management. *Pest Management Science, 67*(5), 493–506.

Guglielmini, A. C., Verdu, A. M. C., & Satorre, E. H. (2017). Competitive ability of five common weed species in competition with soybean. *International Journal of Pest Management, 63*(1), 30–36.

Harker, K. N. (2013). Slowing weed evolution with integrated weed management. *Canadian Journal of Plant Science, 93*(5), 759–764.

Heady, H., & Child, R. D. (1999). *Rangeland ecology and management* (pp. 885–902). Boulder, CO: Westview Press.

Hussain, M. I., Gonzalez, L., & Reigosa, M. J. (2011). Allelopathic potential of *Acacia melanoxylon* on the germination and root growth of native species. *Weed Biology and Management, 11*(1), 18–28.

Inderjit, Seastedt, T. R., Callaway, R. M., Pollock, J. L., & Kaur, J. (2008). Allelopathy and plant invasions: Traditional, congeneric, and bio-geographical approaches. *Biological Invasions, 10*(6), 875–890.

Jabran, K., & Farooq, M. (2013). Implications of potential allelopathic crops in agricultural systems. In *Allelopathy* (pp. 349–385). Berlin, Germany: Springer.

Jabran, K., Mahajan, G., Sardana, V., & Chauhan, B. S. (2015). Allelopathy for weed control in agricultural systems. *Crop Protection, 72*, 57–65.

Javed, S., Javaid, A., & Shoaib, A. (2014). Herbicidal activity of some medicinal plants extracts against *Parthenium hysterophorus* L. *Pakistan Journal of Weed Science Research, 20*(3), 279–291.

Khanh, T. D. (2007). Role of allelochemicals for weed management in rice. *Allelopathy Journal, 19*, 85–96.

Khanh, T. D., Chung, M. I., Xuan, T. D., & Tawata, S. (2005). The exploitation of crop allelopathy in sustainable agricultural production. *Journal of Agronomy and Crop Science, 191*(3), 172–184.

Ladhari, A., Omezzine, F., Dellagreca, M., Zarrelli, A., & Haouala, R. (2013). Phytotoxic activity of *Capparis spinosa* L. and its discovered active compounds. *Allelopathy Journal, 32*(2), 175–190.

Mushtaq, W., & Siddiqui, M. B. (2018). Allelopathy in Solanaceae plants. *Journal of Plant Protection Research, 58*(1), 1–7.

Oerke, E. C. (2006). Crop losses to pests. *The Journal of Agricultural Science, 144*(1), 31–43.

Pan, L., Li, X. Z., Yan, Z. Q., Guo, H. R., & Qin, B. (2015). Phytotoxicity of umbelliferone and its analogs: Structure–activity relationships and action mechanisms. *Plant Physiology and Biochemistry, 97*, 272–277.

Pimentel, D., McNair, S., Janecka, J., Wightman, J., Simmonds, C., O'connell, C., & Tsomondo, T. (2001). Economic and environmental threats of alien plant, animal, and microbe invasions. *Agriculture, Ecosystems & Environment, 84*(1), 1–20.

Qasem, J. R., & Foy, C. L. (2001). Weed allelopathy, its ecological impacts and future prospects: A review. *Journal of Crop Production, 4*(2), 43–119.

Singh, H. P., Batish, D. R., & Kohli, R. K. (2003). Allelopathic interactions and allelochemicals: New possibilities for sustainable weed management. *Critical Reviews in Plant Sciences, 22*, 239–311.

Sodaeizadeh, H., Rafieiolhossaini, M., & Van Damme, P. (2010). Herbicidal activity of a medicinal plant, Peganum harmala L., and decomposition dynamics of its phytotoxins in the soil. *Industrial Crops and Products, 31*(2), 385–394.

Tabaglio, V., Marocco, A., & Schulz, M. (2013). Allelopathic cover crop of rye for integrated weed control in sustainable agroecosystems. *Italian Journal of Agronomy, 8*(1), 1–5.

Tesio, F., & Ferrero, A. (2010). Allelopathy, a chance for sustainable weed management. *International Journal of Sustainable Development and World Ecology, 17*(5), 377–389.

Varshney, J. G., & Babu, M. B. B. P. (2008). Future scenario of weed management in India. *Indian Journal of Weed Science, 40*(1), 1–9.

Vyvyan, J. R. (2002). Allelochemicals as leads for new herbicides and agrochemicals. *Tetrahedron, 58*, 1631–1636.

Young, S. L., Pierce, F. J., & Nowak, P. (2014). Introduction: Scope of the problem—Rising costs and demand for environmental safety for weed control. In *Automation: The future of weed control in cropping systems* (pp. 1–8). Dordrecht, The Netherlands: Springer.

Zeng, R. S. (2014). Allelopathy-the solution is indirect. *Journal of Chemical Ecology, 40*(6), 515–516.

Zimdahl, R. L. (2013). *Fundamentals of weed science* (4th ed., p. 664). San Diego, CA: Academic Press.

Chapter 2
History of Allelopathy

Molisch (1937) derived allelopathy from the two Greek words: allelon (which means 'of each other') and pathos (which means 'to suffer'). Allelobiogenesis or allelopathy characterized by the combination of both biotic and abiotic stresses actuated by donor plants on recipient plants. As per the modern literature, the term allelopathy is an organic chemical interceded negative impedance between plants or microorganisms through its direct or indirect influence (De Albuquerque et al., 2011; Rice, 1984; Willis, 2000; Yang et al., 2011).

The marvel of allelopathy has existed for a great many years, for more than 2000 years. The acknowledgement and comprehension of allelopathy have happened just by escalated scientific records in the course of recent decades (Weston, 2005). The earliest remarks of weed and crop allelopathy were recorded by none other than Theophrastus, 'the father of Botany', who in 300 B.C. wrote in his botanical works about how chickpea 'exhausted' the soil and wrecked weeds (Khalid, Ahmad, & Shad, 2002). Willis (1985) called attention to that Theophrastus (300 B.C.) first saw the deleterious impact of cabbage on a vine and endorsed that it is because of odours. Cato the Elder (234–140 B.C.) wrote in his book on how chickpea and barley 'scorch up' corn land. He also mentioned the harmful effects of walnut trees on different plants (Zeng, Mallik, & Luo, 2008). The historical backdrop of allelopathy could be partitioned into 3 periods of its growth (Singh, Batish, & Kohli, 2001).

1. DeCandolle Phase
   The period spanning late eighteenth to early nineteenth century, particularly between 1785 and 1845
2. Pre-Molisch Phase
   The period of the early twentieth century extending from 1900 to 1920, known by the work of Pickering and Scheiner.

© The Author(s), under exclusive license to Springer Nature Switzerland AG 2020
W. Mushtaq et al., *Allelopathy*, SpringerBriefs in Agriculture,
https://doi.org/10.1007/978-3-030-40807-7_2

3. Post-Molisch Phase

1937 onwards, which actually showed progress ever since 1960 (Willis, 1997).

In communities, distinct plant species may associate in a constructive/positive, neutral, or negative way. Positive connection incorporates either obligatory or non-obligatory mutualism. It is only once in a while the living organisms in a community stay neutral. Negative interactions, however, are more common between the organisms. The unfavourable effect of a neighbouring plant in an affiliation is known as interference (Muller, 1969). Putnam and Tang (1986) have classified interference as:

1. Allelospoly

   More commonly called competition, which incorporates consumption of at least one or more resources attained for the growth and progression of living organisms in an association.

2. Allelo-Mediation

   Discerning harbouring of herbivore that might choose to feed on one/some plant species, thus lending an advantage to another(s) (Szczepanski, 1977).

3. Allelopathy

   Allelopathy, the chemical mechanism of interspecific plant interference, characterized by a negative effect on plant performance in the association.

2.1 Methodological Approaches Used in Allelopathic Studies

Allelopathy is a very complex event comprising of an extensive network interconnecting a wide range of ecological and physiological processes (Scognamiglio et al., 2013). Its elucidation is very difficult to be achieved. The identification of allelochemicals can include phytotoxicity determination as the first step; however, it is not sufficient for the determination of an allelopathic interaction. Different methodologies for allelopathic studies have been reported in the literature meeting specific requirements of some aspects. The optimum method setup is of vital significance for the elucidation of allelopathic interactions involving well-designed interlinked bioassays and field tests (Inderjit & Callaway, 2003).

A brief overview of the laboratory and field bioassays methodologies has been previously described by Wu, Pratley, Lemerle, and Haig (2001), Inderjit and Callaway (2003), Inderjit and Nilsen (2003), and the concerned problems are widely reviewed by Inderjit and Dakshini (1995). Conventional methods for identifying and analysing allelochemicals are based on bioassay-guided fractionation, separation methods along with spectroscopic studies provided satisfactory results (Macias, Oliva, Varela, Torres, & Molinillo, 1999). Water is the most widely used solvent because of its ability to mimic natural conditions have been reported in several studies. However, extraction conditions show considerable differences from the natural systems as the methods involved in the release of allelochemicals into the environment depend on the physicochemical characteristics and the plant organ involved.

Plant extracts have compound nature consisting of a mixture of several secondary metabolites. Synergistic or/and additive effects are often reported to occur

(Reigosa, Souto, & Gonz, 1999). Many studies describe the identification of putative allelochemicals directly in mixtures or in partially purified fractions by HPLC (Chon & Kim, 2002; Thi, Lan, Chin, & Kato-Noguchi, 2008) or GC–MS (Bousquet-Melou et al., 2005).

Isolation of active compounds involves the simple bioassay of plant extracts and their partitioning if active which are further tested for final purification of the active ones. In some studies, identification, isolation, and characterization of active compounds are followed by another test for their phytotoxic potential (Chon & Kim, 2002) but is not always suitable. This bioactivity-guided fractionation of extracts helped in the separation of many phytotoxic compounds. The measurement of phytotoxic activity depends on tests designed specifically for the prospect (Table 2.1).

2.2   Reports on Allelopathic Plants from 2010

The availability of modern instrumentation and proper techniques has played a pivotal role in the identification, extraction, and characterization of allelochemicals assisting the possibility for numerous studies conducted on allelopathy. Also, large numbers of studies have been led under natural and controlled conditions. The following table (Table 2.2) describes the list of available reports on the allelopathic potential of weeds in the agroecosystems since 2010.

Table 2.1 Methodology used in allelopathy studies (Scognamiglio et al., 2013)

Method	Method description	Matrix	References
Petri dishes bioassays	Petri dishes lined with filter paper or solid agar medium at the bottom with recipient plant seeds Measurements (germination rate and percentage, shoot and root length) carried out after some days of incubation	Extracts	Chon and Kim (2002), El Marsni et al. (2011), Hao, Wang, Christie, and Li (2007), Oueslati (2003)
		Partially purified fractions	El Marsni et al. (2011)
		Radical exudates	Chon and Kim (2005), Hao et al. (2007)
		Essential oils	Silva et al. (2012)
		Pure compounds	Chon and Kim (2002), Reigosa et al. (1999), Scognamiglio et al. (2012)
	Soil tested directly	Soil allelopathy	Herranz, Ferrandis, Copete, Duro, and Zalacaín (2006)
	Donor and receiving plant sown together	Direct plant–plant interaction	San Emeterio, Arroyo, and Canals (2004)
Wheat coleoptile assay	Quantification of the wheat apical zone elongation in a liquid medium in the presence of allelochemicals or fractions	Extracts, partially purified fractions, or pure compounds	El Marsni et al. (2011)
Hydroponic and in pot tests	Like petri dishes assays but with older plants Morphological changes and physiological responses measured	Extracts, partially purified fractions or pure compounds	El Marsni et al. (2011), Hussain and Reigosa (2011)
	Isolation of root exudates	Root exudates	Esmaeili, Heidarzade, and Esmaeili (2012)
CRETS (continuous root exudates trapping system)	Seedlings grown in hydroponic nutrient solution on line with a column containing XAD-4 resin, which traps allelochemicals exudate from roots	Root exudates	Hao et al. (2007)
Plant box method	Donor plant root put on one corner of a plant box filled with agar and test seeds placed on agar gel substrate, in order to study the effects of root exudates on seed germination and seedling growth	Root exudates	Fujii, Parvez, Parvez, Ohmae, and Iida (2003)

(continued)

Table 2.1 (continued)

Method	Method description	Matrix	References
Box growth method paired with root image analysis system	Use of the digital camera technology to measure root length, spread, and surface (dynamics) of plants growing in a glass container	Co-growth of donor and receiving plants	Mardani and Yousefi (2012)
Dish pack method	Plant material put into one hole of a 6-well-multi-dish, other holes with test seeds. Seed germination and growth measured. Volatiles analysed by GC–MS	Volatile allelochemicals	Fujii, Matsuyama, Hiradate, and Shimozawa (2005)
Plant sandwich method	Inclusion of plant material into two layers of agar medium (or quartz) used to grow the test plant	Leachates from litter or decomposing plant material	Fujii et al. (2003), Morikawa et al. (2011)
Agar solid tissue culture method	Donor plant powder mixed with agar	Leachates from litter or decomposing plant material	Zuo, Ma, and Ye (2012)
ECAM (equal compartment agar medium)	Donor seeds sown on agar surface in one-half of a glass beaker prefilled agar After the growth, seeds of receiver-weed species sown on the other half of the agar	Co-growth of donor and receiving plants	Wu, Pratley, Lemerle, and Haig (2000)
Greenhouse pot bioassays (or in field)	Plant material spread over soil surface	Leachates from litter or decomposing plant material	Thi et al. (2008)
	Amendment of plant residues	Leachates from litter or decomposing plant material	Matloob, Khaliq, Farooq, and Cheema (2010)
	Amendment of pure compounds	Pure compounds	Bertin, Harmon, Akaogi, Weidenhamer, and Weston (2009)
	Co-growth of donor and receiving plant	Interaction of donor/receiving plant	Labbafi, Hejazi, Maighany, Khalaj, and Mehrafarin (2010)
	Field studies	Various plant extracts	Akemo, Regnier, and Bennett (2000)
Allelochemical soil static concentration determination	Solvent extractions of chemical compounds from the soil matrix paired with HPLC analysis	Allelochemicals in the soil	Scognamiglio et al. (2012)

(continued)

Table 2.1 (continued)

Method	Method description	Matrix	References
Solid-phase root zone extraction (SPRE) method	Use of sorbent material probes placed in the soil	Allelochemicals in the soil	Weidenhamer, Boes, and Wilcox (2009)
In situ silicone tube microextraction method	Use of sorbent material probes placed in the soil	Fate of allelochemicals in the soil	Mohney et al. (2009)
Rhizosphere soil method	Modified sandwich method: The soil surrounding roots used for inclusion in agar	Allelochemicals in the soil	Fujii, Furubayashi, and Hiradate (2005)
Allelochemical solid-phase microextraction (SPME) method	SPME fibre inserted into the stem of the test plant and the adsorbed compounds can be analysed by GC or HPLC	Allelochemical uptake	Loi, Solar, and Weidenhamer (2008)
Radiolabeled allelochemicals	Observing the compound fate within the plant	Allelochemical uptake	Chiapusio, Pellissier, and Gallet (2004)
Flow cytometry	Detection of effects on cell cycle	Allelochemical mode of action	Zhang, Gu, Shi, Zhou, and Yu (2010)
Confocal microscopy	Detection of effects on target cells	Allelochemical mode of action	Chaimovitsh et al. (2010)
Physiology measurements	Measurement of several parameters	Allelochemical mode of action	Hussain and Reigosa (2011)
Use of silenced plants	Plant ability to synthesize and release specific modified	Allelochemical production	Inderjit, Dahl, and Baldwin (2009)
Metabolomics approach	Determination of plant metabolome Plant extract tested on a test species Test species metabolomic analysis	Allelochemical effect on test species and putative mode of action	Scognamiglio (2011)

Table 2.2 List of plants exhibiting allelopathic effects on other plants (Reports from 2010)

Source plant	Target plant	Parts used and its effects	References
Achillea santolina L.	*Vicia faba* L. and *Hordeum vulgare* L.	Aqueous shoot extracts decreased chlorophyll a and b levels, increased carotenoid content and activity of CAT, GPX, SOD, and GR, whereas residues and their aqueous extracts inhibited seedling length	Darier and Tammam (2012)
Ageratum conyzoides L.	*Raphanus sativus* L.	Phytotoxicity of belowground residues changes during decomposition and was reduced upon the addition of soil to the residues	Kaur et al. (2012)
Ageratum conyzoides L.	*Vigna radiata* (L.) R. Wilczek and *Vigna mungo* (L.) Hepper	Various concentrations of whole plant extracts gradually reduced the germination percentage, seedling length, dry biomass, photosynthetic pigments, protein and amino acid contents	Jayaraman and Ramalingam (2014)
Ageratum conyzoides L. and *Cleome viscosa* L.	*Sesamum indicum* L.	Inhibitory effect on germination percentage, root growth, shoot growth, fresh and dry biomass	Natarajan et al. (2014)
Alhagi maurorum L. and *Cardaria draba* L.	*Triticum aestivum* L.	Dry powder of shoots reduces the mineral nutrient concentration of wheat and suppress its growth	Mohammadkhani and Servati (2018)
Amaranthus hybridus L.	*Phaseolus vulgaris* L.	Aqueous extracts affected relative water content (RWC), both vegetative growth and grain yield	Amini and Ghanepour (2013)
Aristolochia esperanzae Kuntze	*Sesamum indicum* L.	Extracts caused marked changes in germination, seedling growth, and 50% reduction in the size of root xylem cells and marked changes in the primary root and in the number of secondary roots	Gatti et al. (2010)

(continued)

Table 2.2 (continued)

Source plant	Target plant	Parts used and its effects	References
Artemisia dubia Wall.	*Parthenium hysterophorus* L.	Aqueous leachates and solvent extract of the whole plant inhibited the germination	Sharma and Devkota (2018)
Atriplex halimus, Atriplex canescens, and *Atriplex nummularia*	*Lactuca sativa* L.	Shoot and root extract inhibited seed germination	Bouchikh-Boucif et al. (2014)
Avena fatua L. and *Secale cereale* L.	*Triticum aestivum* L.	Plant debris and mulch management has a significant effect on germination parameters	Amoghein et al. (2013)
Bothriochloa laguroides var. *laguroides* (DC.) Herter	*Lactuca sativa* L., *Zea mays* L., *Paspalum guenoarum* Arechav., and *Eragrostis curvula* (Schrad.) Nees	Stem and leaf extracts caused inhibition of root and shoot elongation in all four species tested. Aqueous extracts were generally less inhibitory to seed germination	Scrivanti (2010)
Calotropis procera (Ait.) R. Br.	*Triticum aestivum, Raphanus sativus* L., and *Brassica napus* L.	Leaf and flower aqueous extract resulted in delayed germination	Abdel–Farid et al. (2013)
Calotropis procera (Ait.) R. Br.	*Glycine max* L. Merrill.	Germination, growth, coefficient velocity, plumule length, number of leaves were significantly inhibited by aqueous leaf leachates	Ayeni and Akinyede (2014)
Calotropis procera (Ait.) R. Br.	*Ageratum conyzoides* L. *Cannabis sativa* L. and *Trifolium repens* L.	Germination percentage, root length, and shoot length of weed species decreased progressively when treated with increasing conc. of aqueous leaf extract	Gulzar et al. (2014a)
Calotropis procera (Ait.) R. Br.	*Zea mays* L.	Germination and seedling growth were enhanced by leaf extract	Naz and Bano (2013)
Calotropis procera (Ait.) R. Br.	*Brassica oleracea* var. *botrytis*	The leaf, fruit, and flower extract of *C. procera* significantly reduced germination percentage, radicle length, plumule length, dry matter accumulation, and relative water content of the *brassica* seedlings as compared to control	Gulzar and Siddiqui (2015a)

(continued)

Table 2.2   (continued)

Source plant	Target plant	Parts used and its effects	References
Calotropis procera (Ait.) R. Br.	*Allium cepa* L. and *Cassia sophera* (L.)	The leaf aqueous extract significantly reduced the root length, shoot length, and dry biomass of *C. sophera* and the treated young leaves have shrinking and contraction of epidermal cells along with the formation of major grooves, canals, and cyst-like structures and mitotic aberrations in onion root tip cells	Gulzar et al. (2016)
Cannabis sativa L.	*Lactuca sativa* L.	Extracts made of shoot and root parts showed significant effects on germination indices and seedling growth	Mahmoodzadeh et al. (2015)
Capparis spinosa L.	*Lactuca sativa, Raphanus sativus, Silybum marianum,* and *Peganum harmala*	Leaves, stems, and roots extracts were phytotoxic to germination and growth	Ladhari, Omezzine, Dellagreca, Zarrelli, and Haouala (2013)
Cassia sophera (L.) Roxb	*Chenopodium album* L., *Melilotus alba* Medik, and *Nicotiana plumbaginifolia* Viv.	Aqueous extracts significantly reduced the germination percentage, seedling growth, dry biomass, leaf area, relative water content, total protein and chlorophyll content	Gulzar et al. (2014b)
Cassia tora L.	*Brassica juncea* (L.) Coss	Aqueous extract from root, stem, and leaf reduced seed germination, root length, shoot length, chlorophyll content, fresh weight (FW), dry weight (DW), and relative water content (RWC)	Sarkar et al. (2012)
Celosia argentea L.	*Lens culanaris* Medic.	Aqueous leachates of stem, leaves, and inflorescence (flower) increased α -amylase activity	Kengar and Patil (2018)

(continued)

Table 2.2 (continued)

Source plant	Target plant	Parts used and its effects	References
Chenopodium album L.	*Triticum aestivum* L.	Concentrated leaf extracts had detrimental effects on plant height, yield, number of tillers, and spike length	Majeed et al. (2012)
Chenopodium album L., *Amaranthus retroflexus* L., and *Cynodon dactylon* L.	*Carthamus tinctorius* L.	Weed extracts significantly decreased plant height and root dry weight	Rezaie and Yarnia (2009)
Chenopodium murale L. and *Malva parviflora* L.	*Hordeum vulgare* L.	A clear effect of the extract was recorded on the growth parameters, plant height, number of leaves, number of tillers, root fresh and dry weight	Al-johani et al. (2012)
Chromolaena odorata (L.) King and Robinson and *Mikania micrantha* Kunth	*Ageratum conyzoides* L., *Eleusine indica* (L.) Gaertn., and *Cyperus iria* L.	Aqueous leaf extract and leaf debris significantly reduced all seedling growth parameters	Sahid and Yusoff (2014)
Cronopus didymus L. Sm.	*Triticum aestivum* L.	Variable phytotoxicity was exhibited by different extract sources and leaf extract caused the greatest inhibition	Khaliq et al. (2013)
Cronopus didymus L. Sm.	*Oryza sativa* L.	Emergence, seedling growth, and chlorophyll content decreased with increasing concentration of residue amendation in soil	Khaliq et al. (2014)
Cymbopogon nardus (L.) Rendle	*Medicago sativa* L., *Lepidium sativum* L., *Lactuca sativa* L., *Echinochloa crus-galli* (L.) P. Beauv., *Lolium multiflorum* Lam., and *Echinochloa colonum* L.	Inhibitory activity of leaf and root extracts was more inhibitory than stalk extract	Suwitchayanon et al. (2013)
Cynanchum acutum L.	*Hordeum vulgare* L.	Germination percentage, root and shoot length reduced at increasing concentration of watery distillate	Golzardi et al. (2015)

(continued)

Table 2.2 (continued)

Source plant	Target plant	Parts used and its effects	References
Cynanchum acutum (L.) var. Karaj and *Cynanchum acutum* (L.) var. Kerman	Zea mays L.	Watery distillate of leaf stalk and root inhibited seedling length and germination at increasing concentration	Golzardi et al. (2014)
Cynodon dactylon L	*Zea mays* L.	Seed germination and plant growth delayed at the higher concentrations	Bibak and Jalali (2016)
Drimys brasiliensis Miers	*Panicum maximum* and *Euphorbia heterophylla*	Hexane and ethyl acetate fractions of the root extracts showed inhibitory potential on the germination and growth	Anese et al. (2015)
Echinochloa colona L. and *Cyperus iria* L.	*Oryza sativa* L. and Glycine max L.	Organic extracts of root and aerial parts inhibited the seedling growth and germination in a concentration-dependent manner	Chopra et al. (2017)
Echinochloa colona L., *Cleome viscosa* L., and *Ammannia baccifera* L.	*Vigna radiata* (L.) Wilczek	Degree of reduction percentage of all the growth parameters (germination, seedling growth, dry weight) was concentration-dependent	Manikandan and Prabhakaran (2014)
Echinochloa crus-galli (L.) Beauv. and *Leptochloa fusca* (L.) Kunth var. fascicularis (lam.) N. snow	*Oryza sativa* L.	Field plots infested with allelopathic weeds reduced yield, with *Echinochloa* proving more inhibitory than the other weed species	Gealy et al. (2013)
Eclipta alba (L.) Hassk	*Arachis hypogaea* L. and *Vigna radiata* L.	Rhizosphere soil significantly reduced the germination percentage, seedling growth, and dry biomass depending upon the species sensitivity	Gulzar and Siddiqui (2015b)
Erythroxylum monogynum Roxb.	*Solanum lycopersicum* Mill. var. PKM-1	Seed germination, shoot length, root length, fresh weight, and dry biomass were notably decreased by leaf and stem extracts	Alagesaboopathi (2014)
Eupatorium adenophorum Spreng., *Ageratum conyzoides* L., and Lantana camara L.	*Triticum aestivum* cv. HPW-42, *Oryza sativa* cv. Hasanshrasativa I Basmati and *Zea mays* cv. Girija, *Oryza sativa* L.	Incorporation of weed residue in soil had inhibitory effect on percent germination, shoot length, and physiology	Katoch et al. (2012)

(continued)

Table 2.2 (continued)

Source plant	Target plant	Parts used and its effects	References
Eupatorium odoratum L.	*Cicer arietinum* L. and *Cajanus cajana*	Increase in concentration of leaf extract decreased the amylase activity of *Cicer* at all soaking periods and in case of *Cajanus*, increase in concentration as well as seed soaking periods decreases the amylase activity	Madane and Patil (2017)
Euphorbia guyoniana Boiss. and Reut.	*Bromus tectorum* L., *Melilotus indica* (L.) All., and *Triticum aestivum* L.	Germination efficiency, plumule and radicle length reduced upon exposure to aqueous extract	Nasrine et al. (2013)
Euphorbia helioscopia L.	*Triticum aestivum* L., *Cicer arietinum* L., and *Lens culinaris* Medic.	Seedling emergence, seedling vigour index, and total dry weight were significantly reduced upon exposure to rhizosphere soil and aqueous extract of various organs	Tanveer et al. (2010)
Euphorbia himalayensis (Klotzsch) Boiss.	*Triticum aestivum* L., *Lactuca sativa* L., *Poa annua* L., *Festuca rubra* L., and *Trifolium pratense* L.	Root exudates showed allelopathic influence	Liu et al. (2016)
Hyptis suaveolens (L.) Poit.	*Lactuca sativa* L., *Lolium multiflorum* L., and *Echinochloa crus-galli* (L) P. Beauv.	Purified suaveolic acid from the donor plant inhibited the shoot growth	Islam et al. (2014)
Inula crithmoides L.			
Lantana camara L.	*Achyranthes aspera* and *Albizia lebbeck*	Rhizosphere soil and litter reduced density, biomass, diversity indices, richness, and evenness	Singh et al. (2012)
Nicotiana plumbaginifolia Viv.	*Zea mays* L. cv. Uttam	The amount of chlorophyll a, chlorophyll b, carotenoids, protein and nitrate reductase activity decreased upon exposure to leaf and stem aqueous leachate	Singh et al. (2009a)

(continued)

Table 2.2 (continued)

Source plant	Target plant	Parts used and its effects	References
Nicotiana plumbaginifolia Viv.	*Zea mays* L. cv. Uttam	Upon exposure to whole plant extract, leaf water status, chlorophyll content, protein content, nitrate reductase activity (NRA), and activities of antioxidant enzymes decreased; however, proline content increased	Singh et al. (2009b)
Nicotiana plumbaginifolia Viv.	*Helianthus annuus* cv. PAC-36	Leaf, stem, and flower aqueous leachates inhibited germination and seedling growth. Leachates also caused oxidative stress and stimulated the activities of superoxide dismutase (SOD) and catalase (CAT)	Singh et al. (2015)
Ocimum tenuiflorum L.	*Lepidium sativum* L., *Lactuca sativa* L., *Medicago sativa* L., *Lolium multiflorum* Lam., *Echinochloa crus-galli* (L) P. Beauv and *Phleum pretense* L.	Plant extracts reduced significantly the total germination (percent, index, rate and energy), speed of emergence, seedling vigour in addition to root and shoot growth	Islam and Kato-Noguchi (2014)
Parthenium hysterophorus L.	*Brassica juncea*, *Brassica nigra*, *Coriandrum sativum*, *Beta vulgaris*, *Daucus carota*, *Raphanus sativus*, and *Trigonella foenum-graecum*	The root extract inhibited seed germination and the root exudate increases the contents of organic carbon and nitrogen in the rhizosphere soil	Mawal et al. (2015)
Phalaris aquatica L.	*Chloris truncata* R. Br., *Trifolium subterraneum* L., *Medicago trunculata* Gaertn., and *Phalaris aquatica* L.	Aqueous extract neither inhibited germination nor impacted on radicle length however exhibited autotoxicity by inhibiting radicle length	Adams et al. (2010)
Pluchea dioscoridis (L.) DC.	*Corchorus olitorius* L., *Lepidium sativum* L., and *Cynodon dactylon* (L.) Pers	Rhizosphere soil underneath the donor weed caused significant growth reductions	Fahmy et al. (2012)
Poa annua L., *Imperata cylindrica* (L.) Beauv., *Cirsium arvense* (L.) Scop., *Datura alba* Nees, and *Phragmites australis* (Cav.) Steud.	*Zea mays* L., *Avena fatua* L., *Convolvulus arvensis* L., *Ammi visnaga* L., *Rumex crispus* L., and *Asphodelus tenuifolius* Cav.	Germination, shoot length, and shoot weight reduced. Differential reactions were recorded for different weed extracts	Khan et al. (2011a)

(continued)

Table 2.2 (continued)

Source plant	Target plant	Parts used and its effects	References
Rhazya stricta Decne.	*Pennisetum typhoides*	Aqueous extracts showed inhibitory effects on germination and seminal root numbers, while leaf extract significantly decreased the seedling growth	Khan et al. (2011b)
Salvia moorcroftiana, Verbascum thapsus, and *Chenopodium album*	*Triticum aestivum, Helianthus annuus, Avena fatua,* and *Euphorbia helioscopia*	Aqueous extract affected the radicle and plumule growth of seedling and seed germination	Arafat et al. (2015)
Salvia plebeia R. Brown	*Zea mays* var. 30–25 hybrid, *Triticum aestivum* var. Pirsabak-04 and *Sorghum bicolor* L.	Water extract intensely affected the germination, plumule growth, radical growth, chlorophyll content, and fresh and dry weights	Husna et al. (2016)
Setaria viridis L.	*Triticum aestivum* L., *Sorghum vulgare* Pers., *Vigna mungo* (L.) Hepper, *Cucumis sativus* L., *Raphanus sativus* L., and *Galium aparine*	Aqueous extract from aerial parts significantly inhibited seed germination and seedling growth of all bioassay plants except mung bean	Zhu et al. (2013)
Sonchus oleraceus L.	*Trifolium alexandrinum* L., *Brassica nigra* (L.) W.D.J. Koch, *Chenopodium murale* L., *Melilotus indicus* (L.) All., and *Sonchus oleraceus* L.	Plant extract partially inhibited germination and seedling growth at a lower concentration; however, there was complete inhibition of the same parameters at higher concentrations	Gomaa and Abdelgawad (2012)
Tinospora cordifolia (Willd.) Miers.	*Chenopodium album* L., *Chenopodium murale* L., *Cassia tora* L., and *Cassia sophera* L.	Aqueous extracts from root and aerial parts significantly inhibited not only germination and seedling growth but also reduced dry biomass	Raoof and Siddiqui (2012)
Tinospora tuberculata	*Oryza sativa, Solanum lycopersicum, Lactuca sativa,* and *Daucus carota*	Aerial parts affected seedling growth and germination	Aslani et al. (2014)
Vitex negundo L. and *Ricinus communis* L.	*Hyptis suaveolens* L. (Poit.)	The aqueous leaf leachate significantly inhibited the seed germination and early stages of growth	Ramgunde and Chaturvedi (2016)

References

Abdel-Farid, I., El-Sayed, M., & Mohamed, E. (2013). Allelopathic Potential of Calotropis procera and Morettia philaeana. *International Journal of Agriculture and Biology, 15*(1), 130–134.

Adams, A. A., Raman, A., & Nicol, H. I. (2010). Assessment of allelopathic effects of Phalaris aquatica on Chloris truncata, Trifolium subterraneum, Medicago trunculata, and P. aquatica. *Journal of Applied Botany and Food Quality, 83*(2), 163–169.

Akemo, M. C., Regnier, E. E., & Bennett, M. A. (2000). Weed suppression in spring-sown rye (Secale cereale)–pea (Pisum sativum) cover crop mixes. *Weed Technology, 14*(3), 545–549.

Alagesaboopathi, C. (2014). Allelopathic effect of aqueous extract of Erythroxylum monogynum Roxb. on germination and growth of Solanum lycopersicm Mill. Var. PKM-1. *International Journal of Science and Research, 3*(8), 1091–1094.

Al-Johani, N. S., Aytah, A. A., & Boutraa, T. (2012). Allelopathic impact of two weeds, Chenopodium murale and Malva parviflora on growth and photosynthesis of barley (Hordeum vulgare L.). *Pakistan Journal of Botany, 44*(6), 1865–1872.

Amini, R., & Ghanepour, S. (2013). Growth and yield of different types of dry bean affected by smooth amaranth (Amaranthus hybridus L.) shoot extracts. *International Journal of Agriculture and Crop Sciences, 5*(2), 115–119.

Amoghein, M. B., Amoghein, R. S., Tobeh, A., & Jamaati-e-Somarin, S. (2013). Allelopathic effects of extracts and plant residues of wild oat (Avena fatua) and rye (Secale cereale L.) on some germination parameters of wheat crop (Triticum aestivum) in the greenhouse condition. *International Research Journal of Applied and Basic Sciences, 4*(8), 2313–2321.

Anese, S., Gualtieri, S. C. J., Grisi, P. U., Jatoba, L. D. J., & Arduin, M. (2015). Phytotoxic potential of Drimys brasiliensis Miers for use in weed control. Acta Scientiarum. *Agronomy, 37*(4), 505–516.

Arafat, Y., Shahida, K., Lin, W., Fang, C., Sadia, S., Ali, N., & Azeem, S. (2015). Allelopathic evaluation of selected plants extract against broad and narrow leaves weeds and their associated crops. *Academia Journal of Agricultural Research, 3*(10), 226–234.

Aslani, F., Juraimi, A. S., Ahmad-Hamdani, M. S., Omar, D., Alam, M. A., Hashemi, F. S. G. and Uddin, M. K. (2014). Allelopathic effect of methanol extracts from Tinospora tuberculata on selected crops and rice weeds. *Acta Agriculturae Scandinavica, Section B–Soil and Plant Science, 64*(2), 165–177.

Ayeni, M., & Akinyede, O. (2014). Effects of Calotropis procera (Ait.) R. Br. leaves on the germination and early growth of soybeans (Glycine max (L) Merrill). *IOSR Journal of Agriculture and Veterinary Science, 7*(4), 5–9.

Bertin, C., Harmon, R., Akaogi, M., Weidenhamer, J. D., & Weston, L. A. (2009). Assessment of the phytotoxic potential of m-tyrosine in laboratory soil bioassays. *Journal of Chemical Ecology, 35*(11), 1288.

Bibak, H., & Jalali, M. (2016). Allelopathic effects of aqueous extracts of Bermuda grass (Cynodon dactylon L.) on germination, characteristics and seedling growth of corn (Zea maize L.). *Agriculture Science Developments, 5*(2), 11–13.

Bouchikh-Boucif, Y., Labani, A., Benabdeli, K., & Boidielouane, S. (2014). Allelopathic effects of shoot and root extracts from three alien and native Chenopodiaceae species on lettuce seed germination. *Ecologia Balkanica, 6*(2), 51–55.

Bousquet-Melou, A., Louis, S., Robles, C., Greff, S., Dupouyet, S., & Fernandez, C. (2005). Allelopathic potential of *Medicago arborea*, a Mediterranean invasive shrub. *Chemoecology, 15*(4), 193–198.

Chaimovitsh, D., Abu-Abied, M., Belausov, E., Rubin, B., Dudai, N., & Sadot, E. (2010). Microtubules are an intracellular target of the plant terpene citral. *The Plant Journal, 61*(3), 399–408.

Chiapusio, G., Pellissier, F., & Gallet, C. (2004). Uptake and translocation of phytochemical 2-benzoxazolinone (BOA) in radish seeds and seedlings. *Journal of Experimental Botany, 55*(402), 1587–1592.

Chon, S. U., & Kim, J. D. (2002). Biological activity and quantification of suspected allelochemicals from alfalfa plant parts. *Journal of Agronomy and Crop Science, 188*(4), 281–285.

Chopra, N., Tewari, G., Tewari, L. M., Upreti, B., & Pandey, N. (2017). Allelopathic effect of Echinochloa colona L. and Cyperus iria L. weed extracts on the seed germination and seedling growth of rice and soybean. *Advances in Agriculture, 2017*(5748524), 1–5.

Darier, S. M., & Tammam, A. A. (2012). Potentially phytotoxic effect of aqueous extract of Achillea santolina induced oxidative stress on Vicia faba and Hordeum vulgare. *Romanian Journal of Biology-Plant Biology, 57*(1), 3–26.

De Albuquerque, M. B., dos Santos, R. C., Lima, L. M., de Albuquerque Melo Filho, P., Nogueira, R. J. M. C., Da Câmara, C. A. G., & de Rezende Ramos, A. (2011). Allelopathy, an alternative tool to improve cropping systems. A review. *Agronomy for Sustainable Development, 31*(2), 379–395.

El Marsni, Z., Casas, L., Mantell, C., Rodríguez, M., Torres, A., Macias, F. A., & Varela, R. M. (2011). Potential allelopathic of the fractions obtained from sunflower leaves using supercritical carbon dioxide. *The Journal of Supercritical Fluids, 60*, 28–37.

Esmaeili, M., Heidarzade, A., & Esmaeili, F. (2012). Quantifying of common allelochemicals in root exudates of barnyardgrass (*Echinochloa crus-galli* L.) and inhibitory potential against rice (*Oryza sativa*) cultivars. *American-Eurasian Journal of Agricultural & Environmental Sciences, 12*(6), 700–705.

Fahmy, G. M., Al-Sawaf, N. A., Turki, H., & Ali, H. I. (2012). Allelopathic potential of Pluchea dioscoridis (L.) DC. *Journal of Applied Science Research, 8*, 3129–3142.

Fahn, A. (2000). Structure and function of secretory cells. *Advances in Botanical Reseach, 31*, 37–75.

Fujii, Y., Furubayashi, A., & Hiradate, S. (2005). Rhizosphere soil method: A new bioassay to evaluate allelopathy in the field. In J. D. I. Harper, M. An, H. Wu, & J. H. Kent (Eds.), *Proceedings of the 4th world congress on allelopathy establishing the scientific base* (pp. 490–492). Wagga Wagga, NSW: Charles Sturt University.

Fujii, Y., Matsuyama, M., Hiradate, S., & Shimozawa, H. (2005). Dish pack method: A new bioassay for volatile allelopathy. In J. D. I. Harper, M. An, H. Wu, & J. H. Kent (Eds.), *Proceedings of the 4th world congress on allelopathy, "establishing the scientific base"* (pp. 493–497). Wagga Wagga, NSW: Charles Sturt University.

Fujii, Y., Parvez, S. S., Parvez, M. M., Ohmae, Y., & Iida, O. (2003). Screening of 239 medicinal plant species for allelopathic activity using the sandwich method. *Weed Biology and Management, 3*(4), 233–241.

Gatti, A. B., Ferreira, A. G., Arduin, M., & Perez, S. C. G. D. A. (2010). Allelopathic effects of aqueous extracts of Artistolochia esperanzae O. *Kuntze on development of Sesamum indicum L. seedlings. Acta Botanica Brasilica, 24*(2), 454–461.

Gealy, D., Moldenhauer, K., & Duke, S. (2013). Root distribution and potential interactions between allelopathic rice, sprangletop (Leptochloa spp.), and barnyardgrass (Echinochloa crus-galli) based on 13 C isotope discrimination analysis. *Journal of Chemical Ecology, 39*(2), 186–203.

Golzardi, F., Vaziritabar, Y., Vaziritabar, Y., Asilan, K. S., Hasan, M., Sayadi, J., & Sarvaramini, S. (2014). Allelopathic effect of two Cynanchum acutum L. populations on emergence and shoot development of corn. *International Journal of Advanced Life Sciences, 7*(4), 615–627.

Golzardi, F., Vaziritabar, Y., Vaziritabar, Y., Asilan, K. S., Ebadi, S. Z., Sarvaramini, S., & Sayadi, M. H. J. (2015). Allelopathy effect of two Cynanchum acutum L. populations on emergence and shoot development of barley. *Journal of Applied Environmental and Biological Sciences, 5*(1), 166–175.

Gomaa, N. H., & Abdelgawad, H. R. (2012). Phytotoxic effects of *Echinochloa colona* (L.) Link. (Poaceae) extracts on the germination and seedling growth of weeds. *Spanish Journal of Agricultural Research, 10*(2), 492–501.

Gulzar, A., & Siddiqui, M. B. (2015a). Allelopathic effect of Calotropis procera (Ait.) R. Br. on growth and antioxidant activity of Brassica oleracea var. botrytis. *Journal of the Saudi Society of Agricultural Sciences, 16*(4), 375–382.

Gulzar, A., & Siddiqui, M. B. (2015b). Root-mediated allelopathic interference of bhringraj (Eclipta alba L.) Hassk. on peanut (Arachis hypogaea) and mung bean (Vigna radiata). *Applied Soil Ecology, 87*, 72–80.

Gulzar, A., Siddiqui, M. B., & Arerath, U. (2014a). Phytotoxic effects of Calotropis procera (Ait.) R. Br. Extract on three weed plants. *Analele Universitatii Din Oradea Fascicula Biologie, 21*(2), 57–60.

Gulzar, A., Siddiqui, M. B., & Ansari, S. (2014b). Assessment of allelopathic potential of Cassia sophera L. on seedling growth and physiological basis of weed plants. *African Journal of Biotechnology, 13*(9), 037–1047.

Gulzar, A., Siddiqui, M. B., & Ansari, S. (2016). Phenolic acid allelochemicals induced morphological, ultrastructural, and cytological modification on Cassia sophera L. and Allium cepa L. *Protoplasma, 253*(5), 1211–1221.

Hao, Z. P., Wang, Q., Christie, P., & Li, X. L. (2007). Allelopathic potential of watermelon tissues and root exudates. *Scientia Horticulturae, 112*(3), 315–320.

Herranz, J. M., Ferrandis, P., Copete, M. A., Duro, E. M., & Zalacaín, A. (2006). Effect of allelopathic compounds produced by *Cistus ladanifer* on germination of 20 Mediterranean taxa. *Plant Ecology, 184*(2), 259–272.

Husna, Shah, M., Sayyed, A., Shabeena, Aziz, L., Ismail and Gul, H. (2016). Allelopathic effect of Salvia plebia R. Brown on germination and growth of Zea mays var. 30-25 Hybrid, Triticum astivum var. Pirsabak-04 and Sorghum bicolor L. Journal of Applied Environmental and Biological Sciences, 6(4), 93-104.

Hussain, M. I., & Reigosa, M. J. (2011). Allelochemical stress inhibits growth, leaf water relations, PSII photochemistry, non-photochemical fluorescence quenching, and heat energy dissipation in three C3 perennial species. *Journal of Experimental Botany, 62*(13), 4533–4545.

Inderjit, & Callaway, R. M. (2003). Experimental designs for the study of allelopathy. *Plant and Soil, 256*(1), 1–11.

Inderjit, Dahl, C. C. V., & Baldwin, I. T. (2009). Use of silenced plants in allelopathy bioassays: A novel approach. *Planta, 229*(3), 559–575.

Inderjit, & Dakshini, K. M. M. (1995). On laboratory bioassays in allelopathy. *Botanical Review, 61*, 28–44.

Inderjit, & Nilsen, E. T. (2003). Bioassays and field studies for allelopathy in terrestrial plants: Progress and problems. *Critical Reviews in Plant Sciences, 22*(3–4), 221–238.

Islam, A. K. M., & Kato-Noguchi, H. (2014). Allelopathic activity of Leonurus sibiricus on different target plant species. *Journal of Food, Agriculture and Environment, 12*, 286–289.

Islam, A. K. M., Ohno, O., Suenaga, K., & Kato-Noguchi, H. (2014). Suaveolic acid: A potent phytotoxic substance of Hyptis suaveolens. *The Scientific World Journal, 2014*(425942), 1–6.

Jayaraman, P., & Ramalingam, A. (2014). Allelopathy potential of invasive alien species Ageratium conyzoides L.on growth and development of green gram [Vignaradiata (L.) R. Wilczek] and black gram [Vigna mungo (L.) Hepper]. *International Journal of Advances in Pharmacy, Biology and Chemistry, 3*(2), 437–442.

Katoch, R., Singh, A., & Thakur, N. (2012). Effect of weed residues on the physiology of common cereal crops. *Crops, 2*(5), 828–834.

Kaur, S., Singh, H. P., Batish, D. R., & Kohli, R. K. (2012). Phytotoxicity of decomposing belowground residues of Ageratum conyzoides: nature and dynamics of release of phytotoxins. *Acta Physiologiae Plantarum, 34*(3), 1075–1081.

Kengar, Y. D., & Patil, B. J. (2018). Allelopathic influence of Celosia Argentea L. against α-amylase activity in Lens Culanaris Medic. during seed germination. *International Journal for Science and Advance Research in Technology, 4*(1), 420–423.

Khalid, S., Ahmad, T., & Shad, R. A. (2002). Use of allelopathy in agriculture. *Asian Journal of Plant Sciences, 1*(3), 292–297.

Khaliq, A., Hussain, S., Matloob, A., Wahid, A., & Aslam, F. (2013). Aqeous swine cress (Coronopus didymus) extracts inhibit wheat germination and early seedling growth. *International Journal of Agriculture and Biology, 15*(4), 743–748.

Khaliq, A., Hussain, S., Matloob, A., Tanveer, A., & Aslam, F. (2014). Swine cress (Cronopus didymus L. Sm.) residues inhibit rice emergence and early seedling growth. *The Philippine Agricultural Scientist, 96*(4), 419–425.

Khan, M. A., Umm-e-Kalsoom, Khan, M. I., Khan, R., Khan, S. A. (2011a). Screening the alle-lopathic potential of various weeds. Pakistan Journal of Weed Science Research, 17(1), 73-81.

Khan, M., Farrukh, H., & Shahana, M. (2011b). Allelopathic potential of Rhazya stricta Decne. on germination of Pennisetum typhoides. *International Journal of Biosciences, 1*(4), 80–85.

Labbafi, M. R., Hejazi, A., Maighany, F., Khalaj, H., & Mehrafarin, A. (2010). Evaluation of alle-lopathic potential of Iranian wheat (*Triticum aestivum* L.) cultivars against weeds. *Agriculture and Biology Journal of North America, 1*, 355–361.

Ladhari, A., Omezzine, F., Dellagreca, M., Zarrelli, A., & Haouala, R. (2013). Phytotoxic activ-ity of Capparis spinosa L. and its discovered active compounds. *Allelopathy Journal, 32*(2), 175–190.

Liu, X., Tian, F., Tian, Y., Wu, Y., Dong, F., Xu, J., & Zheng, Y. (2016). Isolation and identification of potential allelochemicals from aerial parts of *Avena fatua* L. and their allelopathic effect on wheat. *Journal of Agricultural and Food Chemistry, 64*(18), 3492–3500.

Loi, R. X., Solar, M. C., & Weidenhamer, J. D. (2008). Solid-phase microextraction method for in vivo measurement of allelochemical uptake. *Journal of Chemical Ecology, 34*(1), 70–75.

Macias, F. A., Oliva, R. M., Varela, R. M., Torres, A., & Molinillo, J. M. (1999). Allelochemicals from sunflower leaves cv. Peredovick. *Phytochemistry, 52*(4), 613–621.

Madane, A. N., & Patil, B. J. (2017). Allelopathic effect of Eupatorium odoratum L. on amylase activity during seed germination of Cicer arietinum L. and Cajanuscajan (L) Millsp. *Bioscience Discovery, 8*(1), 82–86.

Mahmoodzadeh, H., Ghasemi, M., & Zanganeh, H. (2015). Allelopathic effect of medicinal plant Cannabis sativa L. on Lactuca sativa L. seed germination. *Acta Agriculturae Slovenica, 105*(2), 233–239.

Majeed, A., Chaudhry, Z., & Muhammad, Z. (2012). Allelopathic assessment of fresh aqueous extracts of Chenopodium album L. for growth and yield of wheat (Triticum aestivum L.). *Pakistan Journal of Botany, 44*(1), 165–167.

Manikandan, V., & Prabhakaran, J. (2014). Allelopathic influence of some weed residues on growth and developmental changes of green gram (Vigna Radiata (L.) Wilczek). *International Journal of Current Biotechnology, 2*(3), 6–10.

Mardani, R., & Yousefi, A. R. (2012). Using image analysis to study the allelopathic potential of wheat cultivars against wild barley (*Hordeum spontaneum*). *International Journal of Applied & Basic Medical Research, 3*, 2281–2288.

Mawal, S. S., Shahnawaz, M., Sangale, M. K., & Ade, A. B. (2015). Assessment of allelopathic potential of the roots of Parthenium hysterophorus L. on some selected crops. *International Journal of Scientific Research in Knowledge, 3*(6), 145–152.

Matloob, A., Khaliq, A., Farooq, M., & Cheema, Z. A. (2010). Quantification of allelopathic potential of different crop residues for the purple nutsedge suppression. *Pakistan Journal of Weed Science Research, 16*(1), 1–12.

Mohammadkhani, N., & Servati, M. (2018). Nutrient concentration in wheat and soil under alle-lopathy treatments. *Journal of Plant Research, 131*(1), 143–155.

Mohney, B. K., Matz, T., LaMoreaux, J., Wilcox, D. S., Gimsing, A. L., Mayer, P., & Weidenhamer, J. D. (2009). In situ silicone tube microextraction: A new method for undisturbed sampling of root-exuded thiophenes from marigold (*Tagetes erecta* L.) in soil. *Journal of Chemical Ecology, 35*(11), 1279.

Molisch, H. (1937). *Der Einfluss einer Pflanze auf die andere-Allelopathic*. Jene, Germany: G. Fischer.

Morikawa, C. I. O., Miyaura, R., Kamo, T., Hiradate, S., Perez, J. A. C., & Fujii, Y. (2011). Isolation of umbelliferone as a principal allelochemical from the Peruvian medicinal plant *Diplostephium foliosissimum* (Asteraceae). *Revista de la Sociedad Quimica del Peru, 77*(4), 285–291.

Muller, C. H. (1969). Allelopathy as a factor in ecological process. *Plant Ecology, 18*(1), 348–357.

Nasrine, S., & El-Taher, S. E. D. H. (2013). Allelopathic effect of Euphorbia guyoniana aqueous extract and their potential uses as natural herbicides. *Sains Malaysiana, 42*(10), 1501–1504.

Natarajan, A., Elavazhagan, P., & Prabhakaran, J. (2014). Allelopathic potential of billy goat weed Ageratum Conyzoides L. and Cleome Viscosa L. on germination and growth of Sesamum Indicum L. *International Journal of Current Biotechnology, 2*(2), 21–24.

Naz, R., & Bano, A. (2013). Effects of Calotropis procera and Citrullus colosynthis on germination and seedling growth of maize. *Allelopathy Journal, 31*(1), 105–116.

Oueslati, O. (2003). Allelopathy in two durum wheat (*Triticum durum* L.) varieties. *Agriculture, Ecosystems & Environment, 96*(1–3), 161–163.

Putnam, A. R., & Tang, C. S. (1986). Allelopathy: State of the science. In A. R. Putnam & C. S. Tang (Eds.), *The science of allelopathy* (pp. 1–19). New York, NY: Wiley.

Ramgunde, V. and Chaturvedi, A. (2016). Allelopathic effect of Ricinus communis L. and Vitex negundo L. on morphological attributes of invasive alien weed: Cassia uniflora Mill. *IRA-International Journal of Applied Sciences, 3*(3), 438–447.

Raoof, K. A., & Siddiqui, M. B. (2012). Allelopathic effect of aqueous extracts of different parts of Tinospora cordifolia (Willd.) Miers on some weed plants. *Journal of Agricultural Extension and Rural Development, 4*(6), 115–119.

Reigosa, M. J., Souto, X. C., & Gonz, L. (1999). Effect of phenolic compounds on the germination of six weeds species. *Plant Growth Regulation, 28*(2), 83–88.

Rezaie, F., & Yarnia, M. (2009). Allelopathic effects of Chenopodium album, Amaranthus retroflexus and Cynodon dactylon on germination and growth of safflower. *Journal of Food, Agriculture and Environment, 7*, 516–521.

Rice, E. L. (1984). *Allelopathy* (2nd ed., p. 421). New York, NY: Academic Press.

Sahid, I., & Yusoff, N. (2014). Allelopathic effects of' Chromolaena odorata '(L.) King and Robinson and Mikania micrantha HBK on three selected weed species. *Australian Journal of Crop Science, 8*(7), 1024.

San Emeterio, L., Arroyo, A., & Canals, R. M. (2004). Allelopathic potential of *Lolium rigidum* Gaud. on the early growth of three associated pasture species. *Grass and Forage Science, 59*(2), 107–112.

Sarkar, E., Chatterjee, S. N., & Chakraborty, P. (2012). Allelopathic effect of Cassia tora on seed germination and growth of mustard. *Turkish Journal of Botany, 36*(5), 488–494.

Scognamiglio, M. (2011). *Phytochemical analysis of Mediterranean plants: Metabolomic approach to study allelopathic interactions among coexisting species* (Ph.D. dissertation, Second University of Naples).

Scognamiglio, M., D'Abrosca, B., Esposito, A., Pacifico, S., Monaco, P., & Fiorentino, A. (2013). Plant growth inhibitors: Allelopathic role or phytotoxic effects? Focus on Mediterranean biomes. *Phytochemistry Reviews, 12*(4), 803–830.

Scognamiglio, M., Esposito, A., D'Abrosca, B., Pacifico, S., Fiumano, V., Tsafantakis, N., & Fiorentino, A. (2012). Isolation, distribution and allelopathic effect of caffeic acid derivatives from *Bellis perennis* L. *Biochemical Systematics and Ecology, 43*, 108–113.

Scognamiglio, M., Fiumano, V., D'Abrosca, B., Pacifico, S., Messere, A., Esposito, A., & Fiorentino, A. (2012). Allelopathic potential of alkylphenols from *Dactylis glomerata* subsp. hispanica (Roth) Nyman. *Phytochemistry Reviews, 5*(1), 206–210.

Scrivanti, L. R. (2010). Allelopathic potential of Bothriochloa laguroides var. laguroides (DC.) Herter (Poaceae: Andropogoneae). *Flora-Morphology, Distribution, Functional Ecology of Plants, 205*(5), 302–305.

Sharma, M., & Devkota, A. (2018). Allelopathic Influences of Artemisia Dubia Wall. *Ex. Besser on Seed Germination and Seedling Vigor of Parthenium Hysterophorus L. Journal of Institute of Science and Technology, 22*(2), 117–128.

Silva, M. P., Piazza, L. A., Lopez, D., Rivilli, M. J. L., Turco, M. D., Cantero, J. J., & Scopel, A. L. (2012). Phytotoxic activity in *Flourensia campestris* and isolation of (−)-hamanasic acid A as its active principle compound. *Phytochemistry, 77*, 140–148.

Singh, H. P., Batish, D. R., & Kohli, R. K. (2001). Allelopathy in agroecosystems: An overview. *Journal of Crop Production, 4920*, 1–41.

Suwitchayanon, P., Pukclai, P., & Kato-Noguchi, H. (2013). Allelopathic activity of Cymbopogon nardus (Poaceae): A preliminary study. *Journal of Plant Studies, 2*(2), 1–6.

Szczepanski, A. J. (1977). Allelopathy as a means of biological control of water weeds. *Aquatic Botany, 3*, 193–197.

Tanveer, A., Rehman, A., Javaid, M. M., Abbas, R. N., Sibtain, M., Ahmad, A. U. H., & Aziz, A. (2010). Allelopathic potential of Euphorbia helioscopia L. against wheat (Triticum aestivum L.), chickpea (Cicer arietinum L.) and lentil (Lens culinaris Medic.). *Turkish Journal of Agriculture and Forestry, 34*(1), 75–81.

Teerarak, M., Laosinwattana, C., & Charoenying, P. (2010). Evaluation of allelopathic, decomposition and cytogenetic activities of Jasminum officinale L. f. var. grandiflorum (L.) Kob. on bioassay plants. *Bioresource Technology, 101*(14), 5677–5684.

Thi, H. L., Lan, P. T. P., Chin, D. V., & Kato-Noguchi, H. (2008). Allelopathic potential of cucumber (Cucumis sativus) on barnyardgrass (*Echinochloa crus-galli*). *Weed Biology and Management, 8*(2), 129–132.

Weidenhamer, J. D., Boes, P. D., & Wilcox, D. S. (2009). Solid-phase root zone extraction (SPRE): A new methodology for measurement of allelochemical dynamics in soil. *Plant and Soil, 322*(1–2), 177–186.

Weston, L. A. (2005). History and current trends in the use of allelopathy for weed management. *Cornell University Turfgrass Times, 13*, 529–534.

Willis, R. J. (1985). The historical bases of the concept of allelopathy. *Journal of the History of Biology, 18*, 71–102.

Willis, R. J. (1997). The history of allelopathy. 2. The second phase (1900 - 1920). The era of S. U. Pickering and the USDA Bureau of Soils. *Allelopathy Journal, 4*, 7–56.

Willis, R. J. (2000). Juglans spp., juglone and allelopathy. *Allelopathy Journal, 7*, 1–55.

Wu, H., Pratley, J., Lemerle, D., & Haig, T. (2000). Evaluation of seedling allelopathy in 453 wheat (*Triticum aestivum*) accessions against annual ryegrass (*Lolium rigidum*) by the equal-compartment-agar method. *Australian Journal of Agricultural Research, 51*(7), 937–944.

Wu, H., Pratley, J., Lemerle, D., & Haig, T. (2001). Allelopathy in wheat (*Triticum aestivum*). *Annals of Applied Biology, 139*(1), 1–9.

Yang, C. Y., Liu, S. J., Zhou, S. W., Wu, H. F., Yu, J. B., & Xia, C. H. (2011). Allelochemical ethyl 2-methyl acetoacetate (EMA) induces oxidative damage and antioxidant responses in Phaeodactylum tricornutum. *Pesticide Biochemistry and Physiology, 100*(1), 93–103.

Zeng, R. S., Mallik, A. U., & Luo, S. M. (2008). *Allelopathy in sustainable agriculture and forestry*. New York, NY: Springer.

Zhang, Y., Gu, M., Shi, K., Zhou, Y. H., & Yu, J. Q. (2010). Effects of aqueous root extracts and hydrophobic root exudates of cucumber (*Cucumis sativus* L.) on nuclei DNA content and expression of cell cycle-related genes in cucumber radicles. *Plant and Soil, 327*(1–2), 455–463.

Zhu, H., Wu, S., Wu, Q., & Peng, C. (2013). Isolation and identification of autotoxic chemicals from Angelica sinensis (Oliv.) Diels. *Journal of Food, Agriculture and Environment, 11*(3/4), 2136–2140.

Zuo, S. P., Ma, Y. Q., & Ye, L. T. (2012). In vitro assessment of allelopathic effects of wheat on potato. *Allelopathy Journal, 30*(1), 1–10.

Chapter 3
Allelopathy Potential of Important Crops

Allelopathy, in simple terms, is a biochemical phenomenon by which a plant influences the growth, germination, and survival of another plant in its vicinity by producing certain chemical inhibitors into the environment known as allelochemicals. Though the use of allelopathic water extracts is economical and environment-friendly yet the reduction in weed biomass is less than herbicides and manual weeding. However, it may be possible to use these allelopathic water extracts with reduced rates of herbicides to increase their efficacy. A number of crops are known to possess allelopathic potential, some of which are enumerated below.

3.1 Rice

Weeds have been a persistent problem for farmers since the advent of agriculture. Rice (*Oryza sativa* L.) is one of the most important crops in Asian countries and a staple food for a majority of the population all over the world. Hence controlling weeds in rice agriculture is of utmost importance. However, rice farmers have one particular advantage; rice has the ability to grow in water, whereas many weeds cannot. Though this ability has been utilized by farmers earlier as a measure to control weeds, growing rice crops in water is a labour-intensive process and irrigation measures are also becoming scarce with time (Olofsdotter, 1998). The extensive use of herbicides for rice cultivation has led to the generation of a wide range of prolific herbicide-resistant weeds (Shibayama, 2001). As a result, extensive research began to exploit the allelopathic potential of rice from the 1970s in countries like the USA, Japan, Korea, India, and China. Some rice varieties release biocidal allelochemicals which might affect major weeds, microbial and pathogenic diversity around rice plants, even soil characteristics (Amb & Ahluwalia, 2016). In 1992, it was proposed

W. Mushtaq et al., *Allelopathy*, SpringerBriefs in Agriculture,
https://doi.org/10.1007/978-3-030-40807-7_3

that rice plants having the following characteristics may have strong allelopathic potential (Garrity, Movillon, & Moody, 1992):

(a) higher yield potential and lower density of some certain weeds than other species,
(b) adequate plant height,
(c) sufficient leaf area.

Certain rice cultivars have a competitive ability against weeds and can suppress weed growth to a large extent. For strongly allelopathic cultivars, allelopathy was the dominant factor determining competitive ability (Olofsdotter, Jensen, & Courtois, 2002). Most of the rice cultivars (under field conditions) express allelopathic potential against barnyard grass, redstem, ducksalad, and monochoria, which represent actual rice–weed interaction, but not against radish and lettuce (Khanh, Xuan, & Chung, 2007). However, in 2010, it was found that many rice varieties exhibited more than 40% inhibition of spinach growth, even by using different varieties (Kabir, Karim, Begum, & Juraimi, 2010). Rice germplasm also exhibited allelopathic potential against ducksalad, redstem, and barnyard grass; out of 3727 accessions from IRRI that are available in the USDA-ARS germplasm resources, more than 500 have shown allelopathic potential against barnyard grass and around 450 have exhibited inhibition on the growth of both redstem and ducksalad (Dilday, Mattice, & Karen, 2001; Olofsdotter, 1998). An important prerequisite for the release of allelopathic germplasm should be a careful evaluation of the environmental consequences of deliberate release of increased quantities of allelochemicals. Such evaluation requires both toxicological and ecotoxicological studies of the allelochemicals responsible for weed suppression (Olofsdotter et al., 2002).

Weeds are known to inherently possess tolerance to certain allelochemicals. Aqueous extracts obtained from the leaves of certain rice cultivars of Malaysia have an inhibitory effect on the seed germination and seedling shoot-root length of such commonly known resistant weed species like *Echinochloa crus-galli, Cyperus difformis, Cyperus iria, Fimbristylis miliacea* (Alam, Hakim, Juraimi, et al., 2018). The allelopathic potential of rice is not limited to the leaves alone but is present in nearly all the plant parts like roots, different stages and colours of hulls and awns. However, a major drawback or limitation observed from such studies the concentration of the allelopathic substances present in the extract or residue mixture maybe much higher than that in the fresh material in the field (Ahn, Chung, & Park, 2000; Jung, Kim, Ahn, et al., 2004). Hence, what is necessary is to isolate the allelochemicals from rice and package them in appropriate concentrations for field application.

Laboratory testing, as well as field testing, has revealed that rice cultivars can suppress both monocot and dicot varieties of weeds, indicating that more than 1 allelochemical is at play for this activity (Olofsdotter, 2001). A large number of compounds such as phenolic acids, fatty acids, indoles, and terpenes have been identified in rice root extracts and decomposing rice residues, as putative allelochemicals that can interact with the surrounding environment. These allelopathic interactions are found to be positive and hence they can be used as an effective contributor for the sustainable and eco-friendly agro-production system (Amb & Ahluwalia, 2016).

A number of allelochemicals were fractionated and isolated from the roots of different rice cultivars viz. azelaic acid, ρ-coumaric acid, 1H-indole-3-carboxaldehyde, 1H-indole-3-carboxylic acid, 1H-indole-5-carboxylic acid, and 1,2-benzenedicarboxylic acid bis (2-ethylhexyl) ester. However, none of these allelochemicals was able to inhibit the growth of barnyard grass alone except ρ-coumaric acid but only at a concentration of >3 mM. Interestingly, at a concentration of around 1 mM, ρ-coumaric acid was able to inhibit the growth of lettuce (Rimando, Olofsdotter, Dayan, & Duke, 2001). This indicates that all these chemicals present in the rice plant work in combination or synergism under field conditions and are able to exhibit and are thus responsible for the high allelopathic potential of the plant.

Improvement of the allelopathic potential of rice cultivars can be achieved through breeding programs. It is possible to improve allelopathy in rice using marker-assisted selection. Optimizing allelopathy in combination with breeding for competitive plant types could result in crop cultivars with superior weed-suppressive qualities (Olofsdotter et al., 2002). The work in allelopathic genetics in rice should go hand-in-hand with the search for better allelochemicals from rice. With increasing knowledge of the function of rice genes, it may soon become possible to identify the most important allelochemicals from physiological function, plant biochemistry in combination with the identified genes.

3.2   Wheat

In order to reduce reliance on chemical pesticides, a lot of research is being carried out in recent years to use allelopathy for increasing crop production with enhanced food quality. In the Kurdistan region of Iraq, where wheat (*Triticum aestivum* L.) is the main field crop, weeds cause around 45–50% of yield loss (Ali, 2013). Wheat, like rice, is another crop that is well known for its allelopathic potential and has been used for the same from a long time. The allelopathic potential of wheat seedlings, residues, straw, and aqueous extracts has been studied on a number of agricultural weeds way back in 1982 (Steinsiek, Oliver, & Collins, 1982); wheat allelopathy has shown reduced infiltration of crops with insects, pests, and diseases. The allelopathic effect of wheat is pronounced to such a large extent that it has consistently reduced re-emergence and growth of a large number of weed species (Wu, Pratley, Lemerle, & Haig, 2001). A number of wheat accessions were screened in the USA for their allelopathic potential; a majority of them were found to reduce the shoot-to-root ratio of vigorously growing weeds like Japanese brome (*Bromus japonicus* L.) and fat hen (*Chenopodium album* L.) by as much as 50% (Spruell, 1984). Like all other allelopathic crops, this biological potential of wheat is influenced by a number of factors like its age, soil pH, carbon and nitrogen content as well as soil water content (Alsaadawi, 2008). Steinsiek et al. (1982) had also concluded that the allelopathic potential of wheat was temperature-dependent; incubation of weed seedlings with wheat extract at 35 °C caused the greatest inhibition on the germination and growth of major weed species like ivy-morning glory, pitted-morning glory, velvetleaf, sicklepod, hemp sesbania, Japanese barnyard millet, etc.

In the last decade, research on wheat allelopathy has increased to a large extent. To reduce the extensive requirements of time, space, and labour, some specific biological screening for the allelopathic potential of wheat under laboratory conditions like extract and seedling screening have been proposed to be carried out before field testing (Wu, An, Liu, et al., 2008). A screening method called equal-compartment-agar method (ECAM) has been employed to assess the allelopathic potential of wheat seedling and it has found to inhibit the growth of a common weed species, annual ryegrass (*Lolium rigidum*) (Wu, Pratley, Lemerle, & Haig, 2000); these wheat accessions inhibited the radicle growth of *L. rigidum* by almost 80% as compared non-wheat controls. Both dicot and monocot weed species cause a nuisance for wheat cultivation; the most prevalent among those are *Secale cereale, Galium tricornutum, Convolvulus arvensis, Avena ludoviciana, Phalaris minor, Vicia villosa,* and *Sinapis arvensis*. Around 4 Iranian wheat cultivars viz. Niknejad, Shiraz, Tabasi, and Roshan were able to significantly reduce the growth of the whole plant as well as root length and germination of these troublesome weed species; C. arvensis was found to be the most sensitive weed to all the cultivars. However, none of them had any effect on the plant dry weight of the weeds (Labbafi, Hejazi, Maighany, et al., 2010). An interesting fact is that the residue from a small amount of any living wheat tissue can inhibit the emergence of several weeds like ivy-leaf morning glory, root pigweed up to as long as 2 months (Alsaadawi, 2008).

The avoidance of allelopathic effects between crops or the exploitation of beneficial interactions in a rotation or a mixed cropping system may have a direct detrimental effect on the crop yield (Rizvi, Haque, Singh, & Rizvi, 1992). The aqueous extracts of wheat residues have shown an allelopathic effect on cotton plants (*Gossypium hirsutum*). Wheat residues release water-soluble allelochemicals that are found to inhibit the growth of corn, sorghum, rye, soybean, etc. Hence, wheat can be suitably used in crop rotation for the increased yield of these crops (Alsaadawi, 2008). Diluted aqueous extracts of roots, leaves, and stems of durum wheat varieties (Karim and Om rabii) have been shown to have allelopathic effect on bread wheat variety 'Ariana' (*Triticum aestivum* L.) and barley 'Manel' (*Hordeum vulgare* L.); leaf extract was found to be the most active (Oueslati, 2003). The application of wheat straw for weed suppression has also been extended to forest plantations; water extracts of wheat straw inhibited propagule growth of the common forest weed red raspberry (*Rubus idaeus* L.) by 44% and this allelopathic effect was further verified in field experiments (Wu et al., 2001).

Often allelopathy is confused with the competition when more than 1 crop is used for weed control. The allelopathic potential of durum wheat *Triticum durum* variety Symmit, barley (*Hordeum vulgare* variety Acsad 14), and oat (*Avena sativa* variety Narimski) toward bread wheat (*Triticum aestivum* variety Aras) was studied in order to differentiate between competition and allelopathy in these plants; it was found that the phytotoxicity of oat was much pronounced, followed by durum wheat, on bread wheat (Ali, 2013). Allelopathy in wheat is quite complex and is often described as dark grey system (grey anatomy); allelopathic potential varied and was discontinuous throughout the wheat life cycle, implying that multiple genes participated in the expression of allelopathy (Lam, Sze, Tong, et al., 2012). Aqueous extracts of decomposing wheat straw contain salts of acetic, propionic, and butyric acids, several phenols, including cinnamic, p-coumaric, ferulic, and sinapic acids. Additionally, ferulic and o-coumaric acids have been identified in the root exudates of wheat (Spruell, 1984). Bread wheat releases DIMBOA which undergoes decomposition to form decomposition product 6-methoxybenzoxazolin-2-one, which has pronounced inhibition on the root growth and seed germination of wild oat, another crop that can act as a serious weed for the survival and productivity of bread wheat (Fragasso, Iannucci, & Papa, 2013).

Breeding programs to improve allelopathy are undertaken widely in rice but not wheat. A hybrid variety of wheat obtained by crossing of a Swedish cultivar having low allelopathic potential and a Tunisian cultivar of high allelopathic potential has the ability to suppress weed growth by approximately 19–20% in fields; however, grain yield of this hybrid variety was reduced by 9%. (Bertholdsson, 2010).Marker studies have revealed 2 major QTLs on chromosome 2B associated with wheat allelopathy. The linkage analysis of genetic markers and the QTLs may improve genetic gains for the allelopathic activity through marker-assisted selection in wheat breeding (Wu, Pratley, Ma, & Haig, 2003). The development of wheat allelopathic cultivars could tremendously reduce the over-reliance on herbicides for weed control. An efficient strategy of biological weed control should consider genetic manipulations to reduce compounds in exudates which stimulate weed growth and rather increase those which inhibit weed growth.

3.3   Sorghum

Knowledge about plant biochemistry, physiology, intra and interplant interactions, the chemistry of natural products, etc. has proved that allelopathic crops along with their allelochemicals can be manipulated for strategical weed management and help to minimize the use of herbicides. Sorghum (*Sorghum bicolor* L.) is an important crop that is well known for its allelopathic potential. It contains a number of allelochemicals (mostly water-soluble) that are phytotoxic to both weeds as well as other crops. According to researchers, sorghum allelopathy can be exploited to yield fruitful results in the field of cropping practices (cover crop, smother crop, companion crop, mixing crop, and smother crop to control weeds and inhibition of nitrification); not only that application of aqueous extracts of sorghum helps to increase crop productivity and lead to weed suppression (Alsaadawi & Dayan, 2009). Like all other allelopathic crops, it is one of the few crops that exhibit allelopathic tolerance, i.e., it has the ability to tolerate and absorb active secretions from other plants (Storozhyk, Mykolayko, & Mykolayko, 2019). During the last 4 decades, extensive work has been done to control weeds using sorghum by the following methods (Alsaadawi & Dayan, 2009):

- Use of sorghum extracts to control weeds.
- Use of sorghum residues as mulch.

- Use of allelopathic crops in crop rotation.
- Use of allelopathic crop in crop mixture and intercropping.

Back in 1982, Lehle et al. reported that 4-week old herbage of sorghum exhibited around 91% inhibitory activity against cumulative cress as compared to root exudates, suggesting that the nature and proportion of allelochemicals in the different plant parts are variable (Lehle & Putnam, 1982). Sudex, a hybrid cultivar, obtained by crossing *Sorghum bicolor* (L.) Moench. and *Sorghum Sudanese* (P.) Stapf., was found to have very high allelopathic potential; among all the plant parts, the sudex shoot tissue exhibited the maximum inhibition on weeds like *Lepidium sativum*, when harvested after 7 days. Partitioning and fractionation, followed by characterization studies revealed that phytoinhibitors were p-hydroxybenzoic acid and p-hydroxybenzaldehyde, potentially the enzymatic breakdown products of the cyanogenic glycoside, dhurrin (Weston, Harmon, & Mueller, 1989). However, dhurrin is toxic to mammalian cells at higher concentrations. The most important allelopathic agent from sorghum was identified as sorgoleone (SGL), a hydroquinone, and phenolic acids; sorghum cultivar, Enkath, was found to have higher concentration of these compounds than Rabeh, when tested against a common companion weed, *Echinochloa colonum* (Ben-Hammouda, Kremer, Minor, & Sarwar, 1995; Sene, Dore, & Pellissier, 2000). In 2007, several allelochemicals were isolated and identified from the sorghum foliage viz. benzoic acid, p-hydroxybenzoic acid, vanillic acid, m-coumaric acid, p-coumaric acid, gallic acid, caffeic acid, ferulic acid, and chlorogenic acid (Cheema, Khaliq, & Farooq, 2007). It is interesting to note that the allelopathic potential in the aqueous extracts of roots, leaves, and decaying residues increased significantly upon exposure to low doses of gamma radiation (Alsaadawi, Al-Uqaili, Al-Hadithy, & Alrubeaa, 1985). It was proposed that allelopathic activity of sorghum was species-specific and depended on source and concentration and investigated that germination and seedling growth of *E. colonum* and radishes were significantly inhibited by allelopathic water extracts of sorghum plant parts (Kim, De Datta, Robles, et al., 1993). In 1986, it was found that weeds such as barnyard grass (*Echinochloa crus-galli* (L.) Beauv.), redroot pigweed (*Amaranthus retroflexus* L.), and red sorrel (*Rumex acetosella* L., when interplanted with sorghum in pots, a significant reduction in their dry weight was observed than weeds grown in monoculture (Panaisuk, Bills, & Leather, 1986). Seed germination of downy brome, a notorious weed in wheat cultivation, was inhibited up to 80% by the root extracts of sorghum but no effect upon treatment with the shoot and leaf extracts (Machado, 2007). Different cultivars of sorghum also differed significantly in their efficacy to suppress the seedling growth of weeds like *Trianthema portulacastrum* and *Cyperus rotundus*; *C. rotundus* being more susceptible to the aqueous leachates of sorghum (Cheema et al., 2007). Water extracts from Medovyi and Dovista, different cultivars of sugar sorghum seeds, at various concentrations (from 5% to 50%) have both a toxic and stimulating effect on germination and germination vigour of sugar beet seeds (Storozhyk et al., 2019). In 2019, an interesting study was carried out on the allelopathic potential of sorghum—stressing sorghum by removing 2 bottom leaves during early growth possibly triggers an increase in the production and exudation of allelochemicals that can inhibit the emergence and growth of the weed, *Amaranthus hybridus*; however, increasing the leaf stripping 4 reduces the amount of photosynthesis carried out by the plant and subsequently reduces production of SGL (Tibugari, Manyeruke, Mafere, et al., 2019).

Solubilization of the allelopathic compounds from sorghum was affected significantly by tilling; allelopathic compounds solubilized rapidly in titled soil due to soil moisture from ample rainfall unlike the non-tilled soil (Purvis, 1990). This difference in the allelopathic effect of sorghum was much pronounced on wheat—tilled sorghum residue often delayed development of the wheat crop but did not affect grain yields, probably because allelopathic compounds degraded in the soil. No-till sorghum stover had little effect on stand establishment but frequently reduced grain yields of wheat, possibly because allelopathic compounds leached into the soil slowly (Roth, Shroyer, & Paulsen, 2000). The future looks bright for further research on the allelopathic properties of sorghum in developing the cropping systems to control weeds and develop sorghum cultivars with superior ability to inhibit weeds using biotechnology techniques.

3.4 Maize

Maize (*Zea mays* L.) is one of the most important cereal crops in the world, particularly in Asian countries like India, Pakistan, China, etc. In addition to its common usage as human food grain, feed for livestock and poultry, corn oil and bakery products, maize provides important and valuable raw materials for agro-industries. Like rice, weed control is one of the major requirements for maize cultivation. The use of herbicides for weed control comes with a large number of harmful; thus, allelopathy offers a great perspective to resolve this critical issue and may be used in different ways to manage weeds (Klein & Miller, 1980).

Maize is well known as an allelopathic plant but has always gained less attention than others like wheat, *Brassica*, rice, etc. (Mahmood, 2009). Back in 1983, Garcia reported that maize or corn plants in fields release allelochemicals through root exudation or rain-leached chemicals that are added to the soil by seeping (Garcia, 1983). These allelochemicals or secondary metabolites are produced in all plant parts of maize; even when corn pollen falls on plants growing nearby (like bottle gourd, watermelon, etc.) their fruiting was found to decline along with scorching of beans (Anaya, Ortega, & Rodriguez, 1992). The decomposition of corn and rye residues produced chemicals in the soil which were observed inhibitory to the growth of lettuce (Chou & Patrick, 1976). The allelopathic effects of the root and leaf extracts of maize increase the mineral content (Ca, P, Na, K, Mg) of rhizospheric soybean soil, promoting healthy growth of soybean plants—this effect is exhibited by both fresh and water-stressed parts of maize plant (Ahmad & Bano, 2013). A similar stimulating effect was exerted by maize aqueous extracts on seed germination, seedling growth, and plant physiology of 3 medicinal plants viz. *Platycodon grandiflorum* ADC, *Scutellaria baicalensis* Georgi, and *Salvia miltiorrhiza* Bge at low concentrations of

0.5–1%. However, at higher concentrations, the extracts exhibit a negative or inhibitory effect on the plants (Peng, 2019). Back in 1962, Guenzi and McCalla had reported that maize and sorghum residues possess highly toxic chemicals at harvest but these become non-toxic after 22–28 weeks of decomposition (Guenzi & McCalla, 1962). In 2010, it was found that extracts from different plant parts of *Zea mays*, together with sorghum, have a pronounced inhibitory effect of the germination and growth of wild barley (Hordeum spontaneum) (Al-tawaha & Odat, 2010).

Fractionation of extracts from maize, followed by spectral analysis, revealed that some of the allelochemicals are 5-chloro-6-methoxy-2-benzoxazolinone (Cl-MBOA) which is a naturally occurring new benzoxazolinone, 6-methoxy-2-benzoxazolinone (MBOA), and 2,4-dihydroxy-1,4-benzoxazin-3-one (DIBOA). At concentrations greater than 0.03 mM for Cl-MBOA and DIBOA and 0.1 mM for MBOA, respectively, the growth of the roots of cress (*Lepidium sativum* L.) seedlings was inhibited; activity of Cl-MBOA was greater than that of its analog, MBOA (Kato-noguchi, 2000; Kato-noguchi, Sakata, Takenokuchi, & Kosemura, 2000). Kato-Noguchi had also reported that the inhibitory activity of 1 allelochemical, DIBOA, was higher in light-grown maize seedlings than those in dark at the same concentration (Kato-noguchi, 1999). Around 18 compounds in the decomposing corn residues were found phytotoxic to seed germination. Among these, salicylaldehyde and butyric, phenylacetic, and 4-phenylbutyric acids were 'volatile' and benzoic, p-hydroxybenzoic, vanillic, ferulic, o-coumarin, o-hydroxyphenylacetic, salicylic, syringic, p-coumaric, trans-cinnamic, and caffeic acids were non-volatile. Resorcinol, p-hydroxybenzaldehyde, and phloroglucinol were also identified (Mahmood, 2009).

References

Ahmad, N., & Bano, A. (2013). Impact of allelopathic potential of maize (*Zea mays* L.) on physiology and growth of soybean (*Glycine maz* (L.) Merr.). *Pakistan Journal of Botany, 45*, 1187–1192.

Ahn, J. K., Chung, I. M., & Park, L. (2000). Allelopathic potential of rice hulls on germination and seedling growth of barnyardgrass. *Agronomy Journal, 92*, 1162–1167.

Alam, A., Hakim, M. A., Juraimi, A. S., Rafii, M. Y., Hasan, M. M., & Aslani, F. (2018). Potential allelopathic effects of rice plant aqueous extracts on germination and seedling growth of some rice field common weeds. *Italian Journal of Agronomy, 13*, 134–140. https://doi.org/10.4081/ija.2018.1066

Ali, K. A. (2013). Allelopathic potential of some crop plant species on bread wheat (*Triticum aestivum*) using equal compartment agar method. *Journal of Agriculture and Veterinary Science, 2*, 52–55.

Alsaadawi, I. S. (2008). Allelopathic influence of decomposing wheat residues in agroecosystems. *Journal of Crop Production, 4*, 185–196. https://doi.org/10.1300/J144v04n02

Alsaadawi, I. S., Al-Uqaili, J. K., Al-Hadithy, S. M., & Alrubeaa, A. J. (1985). Effect of gamma irradiation on allelopathic potential of *Sorghum bicolor* against weeds and nitrification. *Journal of Chemical Ecology, 11*, 1737–1738.

Alsaadawi, I. S., & Dayan, F. E. (2009). Potentials and prospects of sorghum allelopathy in agroecosystems. *Allelopathy Journal, 24*, 255–270.

Al-tawaha, A. R. M., & Odat, N. (2010). Use of sorghum and maize allelopathic properties to inhibit germination and growth of wild barley (*Hordeum spontaneum*). *Notulae Botanicae Horti Agrobotanici Cluj-Napoca, 38*, 124–127.

Amb, M., & Ahluwalia, A. (2016). Allelopathy: Potential role to achieve new milestones in rice cultivation. *Rice Science, 23*, 165–183. https://doi.org/10.1016/j.rsci.2016.06.001

Anaya, A. L., Ortega, R. C., & Rodriguez, V. N. (1992). Impact of allelopathy in the traditional management of agroecosystems in Mexico. In S. H. Rizvi & V. Rizvi (Eds.), *Allelopathy: Basic and applied aspects* (pp. 271–301). London, UK: Chapman and Hall Publication.

Ben-Hammouda, M., Kremer, R. J., Minor, H. C., & Sarwar, M. (1995). A chemical basis for differential allelopathic potential of sorghum hybrids on wheat. *Journal of Chemical Ecology, 21*, 775–786.

Bertholdsson, N.-O. (2010). Breeding spring wheat for improved allelopathic potential. *Weed Research, 50*, 49–57. https://doi.org/10.1111/j.1365-3180.2009.00754.x

Cheema, Z. A., Khaliq, A., & Farooq, M. (2007). Allelopathic potential of sorghum (*Sorghum bicolor* L. Moench) cultivars for weed management. *Allelopathy Journal, 20*, 167–178.

Chou, C., & Patrick, Z. (1976). Identification and phytotoxic activity of compounds produced during decomposition of corn and rye residues in soil. *Journal of Chemical Ecology, 2*, 369–387.

Dilday, R. H., Mattice, J. D., & Karen, A. (2001). Allelopathic potential in rice germplasm against ducksalad, redstem and barnyard grass. *Journal of Crop Production, 4*, 287–301. https://doi.org/10.1300/J144v04n02

Fragasso, M., Iannucci, A., & Papa, R. (2013). Durum wheat and allelopathy: Towards wheat breeding for natural weed management. *Frontiers in Plant Science, 4*, 1–8. https://doi.org/10.3389/fpls.2013.00375

Garcia, A. G. (1983). *Seasonal variation in allelopathic effects of corn residue on corn and cress seedlings*. Ames, IA: Iowa State University.

Garrity, D. P., Movillon, M., & Moody, K. (1992). Differential weed suppression ability in upland rice cultivars. *Agronomy Journal, 82*, 586–591.

Guenzi, W., & McCalla, T. (1962). Inhibition of germination and seedling development by crop residues. *Soil Science Society of America Journal, 26*, 456–458.

Jung, W. S., Kim, K. H., Ahn, J. K., Hahna, S. J., & Chungb, I. M. (2004). Allelopathic potential of rice (*Oryza sativa* L.) residues against *Echinochloa crus-galli*. *Crop Protection, 23*, 211–218. https://doi.org/10.1016/j.cropro.2003.08.019

Kabir, A., Karim, S., Begum, M., & Juraimi, A. S. (2010). Allelopathic potential of rice varieties against spinach (*Spinacia oleracea*). *Journal of Agricultural Biology, 12*, 809–815.

Kato-noguchi, H. (1999). Effect of light-irradiation on allelopathic potential of germinating maize. *Phytochemistry, 52*, 1023–1027.

Kato-noguchi, H. (2000). Allelopathy in Maize II.: Allelopathic potential of a new benzoxazolinone, 5-chloro-6-methoxy-2- benzoxazolinone and its analogue. *Plant Production Science, 3*, 47–50. https://doi.org/10.1626/pps.3.47

Kato-noguchi, H., Sakata, Y., Takenokuchi, K., & Kosemura, S. (2000). Allelopathy in Maize I.: Isolation and identification of allelochemicals in maize seedlings. *Plant Production Science, 3*, 43–46. https://doi.org/10.1626/pps.3.43

Khanh, T. D., Xuan, T. D., & Chung, I. M. (2007). Rice allelopathy and the possibility for weed management. *Annals of Applied Biology, 151*, 325–339. https://doi.org/10.1111/j.1744-7348.2007.00183.x

Kim, S. Y., De Datta, S. K., Robles, R. P., Kim, K. U., Lee, S. C., & Shin, D. H. (1993). Allelopathic effect of sorghum extract and residues on selected crops and weeds. *Korean Journal of Weed Science, 14*, 34–41.

Klein, R. R., & Miller, D. A. (1980). Allelopathy and its role in agriculture. *Communications in Soil Science and Plant Analysis, 11*, 43–56. https://doi.org/10.1080/00103628009367014

Labbafi, M. R., Hejazi, A., Maighany, F., Samari Khalaj, H. R., & Mehrafarin, A. (2010). Evaluation of allelopathic potential of Iranian wheat (*Triticum aestivum* L.) cultivars against weeds. *Agriculture and Biology Journal of North America, 1*, 355–361.

Lam, Y., Sze, C. W., Tong, Y., Ng, T. B., Tang, S. C. W., Ho, J. C. M., ... Zhang, Y. (2012). Research on the allelopathic potential of wheat. *Agricultural Sciences, 3*, 979–985.

Lehle, F. R., & Putnam, A. R. (1982). Quantification of allelopathic potential of sorghum residues by novel indexing of Richards' function fitted to cumulative cress seed germination curves. *Plant Physiology, 69*, 1212–1216.

Machado, S. (2007). Allelopathic potential of various plant species on downy brome: Implications for weed control in wheat production. *Agronomy Journal, 99*, 127–132. https://doi.org/10.2134/agronj2006.0122

Mahmood, A. (2009). *Weed management in maize (Zea mays L.) through allelopathy*. Faisalabad, Pakistan: University of Agriculture Faisalabad.

Olofsdotter, M. (Ed.). (1998). *Allelopathy in rice*. Manila, Philippines: International Rice Research Institute.

Olofsdotter, M. (2001). Rice—A step toward use of allelopathy. *Agronomy Journal, 93*, 3–8.

Olofsdotter, M., Jensen, L., & Courtois, B. (2002). Improving crop competitive ability using allelopathy - An example from rice. *Plant Breeding, 121*, 1–9.

Oueslati, O. (2003). Allelopathy in two durum wheat (*Triticum durum* L.) varieties. *Agriculture, Ecosystems and Environment, 96*, 161–163.

Panaisuk, O., Bills, D. D., & Leather, G. R. (1986). Allelopathic influence of *Sorghum bicolor* on weeds during germination and early development of seedlings. *Journal of Chemical Ecology, 12*, 1533–1543.

Peng, X. (2019). Allelopathic effects of water extracts of maize leaf on three Chinese herbal medicinal plants. *Notulae Botanicae Horti Agrobotanici Cluj-Napoca, 47*, 194–200. https://doi.org/10.15835/nbha47111226

Purvis, C. (1990). Differential responses of wheat to retained crop stubbles: I. Effect of stubble type and degree of decomposition. *Australian Journal of Agricultural Research, 41*, 225–242.

Rimando, A. M., Olofsdotter, M., Dayan, F. E., & Duke, S. O. (2001). Searching for rice allelochemicals: An example of bioassay-guided isolation. *Agronomy Journal, 93*, 16–20.

Rizvi, S. J. H., Haque, H., Singh, V. K., & Rizvi, V. (1992). A discipline called allelopathy. In S. J. H. Rizvi & V. Rizvi (Eds.), *Allelopathy: Basic and applied aspects* (pp. 1–10). Dordrecht, The Netherlands: Springer.

Roth, C. M., Shroyer, J. P., & Paulsen, G. M. (2000). Allelopathy of sorghum on wheat under several tillage systems. *Agriculture, Food and Analytical Bacteriology, 92*, 855–860.

Sene, M., Dore, T., & Pellissier, F. (2000). Effect of phenolic acids in soil under and between rows of a prior sorghum (*Sorghum bicolor*) crop on germination, emergence, and seedling growth of peanut (*Arachis hypogea*). *Journal of Chemical Ecology, 26*, 625–637.

Shibayama, H. (2001). Weeds and weed management in rice production in Japan. *Weed Biology and Management, 1*, 53–60.

Spruell, J. A. (1984). *Allelopathic potential of wheat accessions*. Norman, OK: The University of Oklahoma.

Steinsiek, J. W., Oliver, L. R., & Collins, F. C. (1982). Allelopathic potential of wheat (*Triticum aestivum*) straw on selected weed species. *Weed Science, 30*, 495–497.

Storozhyk, L., Mykolayko, V., & Mykolayko, I. (2019). Allelopathic potential of sugar sorghum (*Sorghum bicolor* (L.) Moench) seeds. *Journal of Microbiology and Biotechnology of Food Science, 9*, 93–98. https://doi.org/10.15414/jmbfs.2019.9.1.93-98

Tibugari, H., Manyeruke, N., Mafere, G., Chakavarika, M., Nyamuzuwe, L., Marumahoko, P., & Mandumbu, R. (2019). Allelopathic effect of stressing sorghum on weed growth. *Cogent Biology, 5*, 1–10. https://doi.org/10.1080/23312025.2019.1684865

Weston, L. A., Harmon, R., & Mueller, S. (1989). Allelopathic potential of sorghum-sudangrass hybrid (Sudex). *Journal of Chemical Ecology, 15*, 1855–1865.

Wu, H., An, M., Liu, D. L., Pratley, J., & Lemerle, D. (2008). Recent advances in wheat allelopathy. In R. Zeng, A. Mallik, & S. Luo (Eds.), *Allelopathy in sustainable agriculture and forestry* (pp. 235–254). New York NY: Springer.

Wu, H., Pratley, H., Lemerle, D., & Haig, T. (2000). Laboratory screening for allelopathic potential of wheat (*Triticum aestivum*) accessions against annual ryegrass (*Lolium rigidum*). *Australian Journal of Agricultural Research, 51*, 259–266.

Wu, H., Pratley, J., Lemerle, D., & Haig, T. (2001). Allelopathy in wheat (*Triticum aestivum*). *Annals of Applied Biology, 139*, 1–9.

Wu, J., Pratley, H., Ma, W., & Haig, T. (2003). Quantitative trait loci and molecular markers associated with wheat allelopathy. *Theoretical and Applied Genetics, 107*, 1477–1481. https://doi.org/10.1007/s00122-003-1394-x

Chapter 4
Allelopathy Potential of Weeds Belonging to the Family

4.1 Asteraceae

Asteraceae consists of 8–10% of angiosperm species, around 1600–1700 genera with 24,000 species (Funk, Susanna, Steussy, & Robinson, 2009). Several plants in the Asteraceae family, especially *chromolaena odorata, Helianthus annuus,* and *Tithonia diversifolia* (of which remnants of *Ageratum conyzoides, Vernonia amygdalina, and Artemisia annua*) are revealed to contain huge amounts of allelochemicals, especially in their leaves, reducing the growth of various plants (Eze & Gill, 1992). Parthenium (*Parthenium hysterophorus* L.), commonly known as congress grass, carrot weed, ragweed Parthenium, is an invasive poisonous herbaceous annual weed of this family. It decreases pasture productivity, affects livestock health, human health and activities, alters ecology and biodiversity, and competes with crop plants for available environmental resources thereby causing enormous yield loss. Parthenin is revealed to be the principal constituent of this plant which accounts for its allelopathy (Mawal, Shahnawaz, Sangale, & Ade, 2015). Netsere (2015) noticed the allelopathic effect of the whole plant of Parthenium on germination and growth of maize and sorghum. Similarly, the powder extract of *Parthenium hysterophorus* reduced the germination rate and growth of wheat plants (Anwar et al., 2016). Sorecha and Bayissa (2017) showed the allelopathic effects of different doses of aqueous extracts of leaf, stem, and root aqueous extracts of *Parthenium* on germination and vegetative growth of peanut and soybean. The allelopathic impact of *Parthenium hysterophorus* extract on the onion meristematic cell influenced mitotic depression leading to chromosomal deformity such as fragments, stickiness nuclear vacuolation, bridge, laggards, and micronuclei. The reduction of DNA content by Parthenium leads to alteration of normal metabolic activity which is a potential threat to genomic balance (Sinha, 2009). Leaf extract of *Parthenium hysterophorus* was reported to inhibit the germination rate and root length of *Cicer arietinum* (Shikha & Jha, 2016). Siyar et al. (2018) reported that the aqueous extract of leaves,

© The Author(s), under exclusive license to Springer Nature Switzerland AG 2020
W. Mushtaq et al., *Allelopathy*, SpringerBriefs in Agriculture,
https://doi.org/10.1007/978-3-030-40807-7_4

stems, and roots of two weeds of Asteraceae (*Artemisia annua* and *Taraxicum officinalis*) had an inhibitory effect on seed germination of wheat and maize, where *Artemisia annua* proved more phytotoxic than *Taraxicum officinalis*. Kamal (2011) observed that the allelochemicals released by sunflower hampered germination and decreased the shoot length, root length, fresh weight, dry weight, chlorophyll contents and also reduced the level of hormones, GA, IAA in wheat. Leachate of *Achillea biebersteinii* was observed to negatively affect percentage and rate of germination and radical and shoot length of pepper. Decreased content of chlorophyll carotenoids and protein was also noticed due to allelochemical stress (Abu-Romman, 2011). Arora, Batish, Singh, and Kohli (2015) showed that *Tagetes minuta* oil retarded seed germination and suppressed seedling growth of invasive weeds—*Chenopodium murale* L., *Phalaris minor* Retz., and *Amaranthus viridis* L by promoting physiological changes that altered chlorophyll content in these plants. Barroso Aragão et al. (2017) spotted the allelopathic effect of *Lepidaploa rufogrisea* extracts through reduction of the root growth and germination speed index in *Lactuca sativa*. Similarly aqueous and methanolic extract of Tithonia diversifolia significantly reduced seed germination, seedling growth, and the biochemical parameters and growth of *Vigna unguiculata* (Oyeniyi, Odekanyin, Kuku, & Otusanya, 2016).

4.2 Convolvulaceae

Convolvulaceae, commonly known as the bindweed or morning glory family, is a family of dicotyledonous flowering plants, comprising of nearly 60 genera with around 1650 species, distributed across the globe. *Cuscuta* (dodder) may be a critical parasite of many crop plants, while *Convolvulus arvensis* and *Calystegia* species (bindweeds) may also be persistent vigorous weeds of gardens and agricultural land. The genus Ipomoea (Convolvulaceae) is composed of about 600 species, and its prolificacy of biologically active secondary metabolites is perhaps its most notable attribute.

The chemical nature of secondary compounds of *Ipomoea cairica* was assessed and two compounds, 3–3′-5-trihidroxy-4′-7-dimethoxyflanove and 3-3′-5-trihidroxy-4′-7-dimethoxyflavone-3-O-sulphate, were found to be accountable for its allelopathic effects on radish (*Raphanus sativus* L.), cucumber (*Cucumis sativus* L.), Chinese cabbage (*Brassica pekinensis*), and a weed *Ligularia virgaurea*, making it potential herbicides on the basis of natural products. A strong effect of leaf extract of *Ipomoea cairica* on *Parthenium hysterophorus* was reported by Srivastava and Shukla (2016). They observed a significant reduction in the total chlorophyll and carotenoid content by the leaf extract. Likewise, chlorophyll 'a' and chlorophyll 'b' contents were also reduced in the weed. Different concentrations of the leaf extracts of *Ipomoea cairica* were revealed to suppress the growth and reduce chlorophyll pigment content of *Parthenium hysterophorus* (Srivastava, Shahi, & Kumar, 2015). Plant extract of Ipomoea carnea decreased the germination and root and shoot elongation in kodo millet (Oudhia, 2000). Jain, Joshi, and Joshi (2017) also reported the allelopathic capability of *Ipomoea carnea* against two common weed plants, i.e., *Cassia fistula* and *Amaranthus spinosus*. It was observed that leaf aqueous extract of *Ipomoea cairica* inhibited root growth in *Allium cepa*. Chromosomal aberrations were also noticed and the mitotic index reduced with increasing concentrations of extracts. The extracts generated the accretion of reactive oxygen species (ROS) (Yuan, Li, Xiong, & Zhang, 2019). Extracts of *Ipomoea batatas* were observed to hamper the germination index and decrease the germination rate in *Amaranthus palmeri* (Hernández Aro, Hernández, & Guillén, 2015). Methanolic extracts of roots, stems, and leaves of *Convolvulus arvensis* were found to decrease the mitotic index and also cause chromosome aberrations and modulate phytohormone levels in *Zea mays* (Sunar & Agar, 2017). Similarly, leachates of *Convolvulus arvensis* reduce the seed germination and seedling growth of wheat (Mishra, 2018).

4.3 Solanaceae

Solanaceae includes about 100 genera and 2500 species. The family Solanaceae comprises numerous significantly economic important plants as vegetables and drug plants. However, some of the plants are found to grow like weeds. Mushtaq and Siddiqui (2018) have reported the comprehensive allelopathic potential of Solanaceae plants. The members with allelopathic potential are found to contain sesquiterpenoid phytoalexins (Elakovich, 1987). *Datura stramonium* has been found to synthesize a number of tropane alkaloids such as atropine, hyoscyamine, and scopolamine, which reduce the growth of associated plant roots and stems. Aqueous extracts of this weed decreased germination, length of plants, and dry weight of soybean plant (Ramona et al., 2018) and other leguminous crops: common bean, cowpea, pigeon pea, and alfalfa (Dafaallah, Mustafa, & Hussein, 2019). Elisante, Tarimo, and Ndakidemi (2013) reported that aqueous leaf and seed extracts of *Datura stramonium* reduced total chlorophyll content and decreased root and shoot length in *Cenchrus ciliaris* and *Neonotonia wightii*. It was observed that aqueous leaf extract of *Datura metel* reduced the germination percentage of Parthenium (Ramachandran & Venkataraman, 2016). Crude ethanol extracts of leaves of *Solanum megalochiton* inhibited seed germination, growth, and respiration of *Lactuca sativa* L. and *Allium cepa* (Krause et al., 2016). Leaf extract of *Withania somnifera* reduced germination and growth of seedling and also showed an inhibitory effect on sugar and protein content of wheat seedlings (Mandal, Mamta, Mandal, Mandal, & Sinha, 2018).

4.4 Verbenaceae

This family commonly referred to as the verbena or vervain family comprises herbs, shrubs, and trees. The family accounts for some 35 genera and 1200 species abundant in tropical regions of the world. Verbenaceae family plants find a wide utilization in the traditional therapeutic systems of several countries. Various plants have been considered to contain bioactive phytochemicals of significant pharmacological effects. There are various reports of allelopathic activity for the Verbenaceae family. Lantana is one of the most invasive plants growing worldwide (Choyal & Sharma, 2011). The extent of its invasiveness has been assigned to its capability for the production of allelopathic chemicals that retard germination, growth, and metabolism of many crops, weeds (Mishra, 2015). Allelopathic impact on seed germination and growth of lantana seedling have been revealed in different species like *Brassica juncea* L., (Ahmed, Uddin, Khan, Mukul, & Hossain, 2007), *Oryza sativa* L. (Hossain & Alam, 2010), *Phaseolus radiatus* (Gantayet, Adhikary, Lenka, & Padhy, 2014). In lantana, the whole part of the plant contains allelochemicals. Monoterpenes sesquiterpenes, flavonoids, iridoid glycoside, furanonaphoquinones, STH steroids, triterpenes, and diterpenes form the main constituents of extracts (Mishra, 2015). It was reported that the methanolic extract of *Lantana camara* caused a remarkable suppression in the germination of seeds and decreased radical and plumule length of specific weeds (like *Rumex dentatus, Avena fatua, Chenopodium album, Euphorbia helioscopia*) of wheat and can be used in further detailed assessment for searching bioherbicide from this plant (Anwar et al., 2018). Similarly, Gantayet et al. (2014) reported allelopathic effects of various concentrations extracted from leaf-litter dust of *Lantana camara* on the green gram (*Phaseolus radiates*). Various concentrations of leaf-litter dust prompted a notable negative effect on the growth and yield of the plants. Rajendiran, Yogeswari, and Arulmozhi (2014) reported that leachates from the *Lantana camara* affected the seedling growth and mitotic index of *Vigna mungo*. Aqueous extracts of this weed were also reported to cause the chromosomal aberrations in the test crop. Aqueous leaf extracts of *Lantana* had an inhibitory effect on the root length and shoot length of *Cicer arietinum* (Shrivastav & Jha, 2016). The allelopathic impact was more noticeable on the root growth of the plants. Enyew and Raja (2015) observed the allelopathic effect of leaf powder of Lantana camara on *Zea mays* and *Triticum turgidum* and a remarkable suppression of seed germination, speed of germination, shoot and root length, and biomass of wheat and maize was observed by these authors. Benzarti, Lahmayer, Dallali, Chouchane, and Hamdi (2016) displayed that aqueous extracts of *Verbena officinalis* L. and *Aloysia citrodora* L. exhibited an allelopathic potential on the seed germination and the radical growth of *Lactuca sativa* and *Phalaris canariensis*. The leaf extract of *Tectona grandis* exhibited a negative effect on seed germination and seedling growth, root vigour index, seedling vigour index, and decrease of biomass production of *Plumbago zeylanica* (Biswas & Das, 2016). *Lippia alba* (Mill.) referred to as 'arbustus lemongrass', 'field lemongrass', 'field rosemary', 'wild rosemary', 'wild grapefruit', '*Melissa* false', and 'brazilian lemongrass' possess a significant allelopathic effect on crops. It was reported that extracts from upper parts of this exhibit decreased growth of hypocotyls and radicles in *Lactuca sativa* and *Allium cepa* (Teixeira de Oliveira et al., 2019).

References

Abu-Romman, S. (2011). Allelopathic potential of *Achillea biebersteinii* Afan. (Asteraceae). *World Applied Sciences Journal, 15*(7), 947–952.

Ahmed, R., Uddin, M. B., Khan, M. A. S. A., Mukul, S. A., & Hossain, M. K. (2007). Allelopathic effects of *Lantana camara* on germination and growth behavior of some agricultural crops in Bangladesh. *Journal of Forest Research, 18*, 301–304.

Anwar, T., Ilyas, N., Qureshi, R., Rahim, B. Z., Maqsood, M., Ansari, K. A., ... Panni, M. K. (2018). Allelopathic potential of *Lantana camara* against selected weeds of wheat crop. *Applied Ecology and Environmental Research, 16*(6), 6741–6760.

Anwar, T., Khalid, S., Saeed, M., Mazhar, R., Qureshi, H., & Rashid, M. (2016). Allelopathic interference of leaf powder and aqueous extracts of hostile weed: *Parthenium hysterophorus* (Asteraceae). *Science International, 4*, 86–93.

Arora, K., Batish, D. R., Singh, H. P., & Kohli, R. K. (2015). Allelopathic potential of the essential oil of wild marigold (*Tagetes minuta* L.) against some invasive weeds. *Journal of Environmental and Agricultural Sciences, 3*, 56–60.

Barroso Aragão, F., Tebaldi Queiroz, V., Ferreira, A., Vidal Costa, A., Fontes Pinheiro, P., Tavares Carrijo, T., & Fonseca Andrade-Vieira, L. (2017). Phytotoxicity and cytotoxicity of *Lepidaploa rufogrisea* (Asteraceae) extracts in the plant model *Lactuca sativa* (Asteraceae). *Revista de Biologia Tropical, 65*(2), 435–443.

Benzarti, S., Lahmayer, I., Dallali, S., Chouchane, W., & Hamdi, H. (2016). Allelopathic and antimicrobial activities of leaf aqueous and methanolic extracts of *Verbena officinalis* L. and *Aloysia citrodora* L.(Verbenaceae): A comparative study. *Med Aromat Plants, 5*, 1–9.

Biswas, K., & Das, A. P. (2016). Allelopathic effects of teak (*Tectona grandis* L.f.) on germination and seedling growth of *Plumbago zeylanica* L. *Pleione, 10*(2), 262–268.

Choyal, R., & Sharma, S. K. (2011). Allelopathic effects of *Lantana camara* (Linn.) on regeneration in *Funaria hygrometrica*. *Indian Journal of Fundamental and Applied Life Science, 1*, 177–182.

Dafaallah, A. B., Mustafa, W. N., & Hussein, Y. H. (2019). Allelopathic effects of jimsonweed (*Datura Stramonium* L.) seed on seed germination and seedling growth of some leguminous crops. *International Journal of Innovative Approaches in Agricultural Research, 3*(2), 321–331.

Elakovich, S. D. (1987). Sesquiterpenes as phytoalexins and allelopathic agents. *Ecology and Metabolism of Plant Lipids, 7*, 93–108.

Elisante, F., Tarimo, M. T., & Ndakidemi, P. A. (2013). Allelopathic effect of seed and leaf aqueous extracts of *Datura stramonium* on leaf chlorophyll content, shoot and root elongation of *Cenchrus ciliaris* and *Neonotonia wightii*. *American Journal of Plant Sciences, 4*(12), 2332.

Enyew, A., & Raja, N. (2015). Allelopathic effect of Lantana camara L. leaf powder on germination and growth behaviour of maize, *Zea mays* Linn. and wheat, *Triticum turgidum* Linn. cultivars. *Asian Journal of Agricultural Science, 7*(1), 4–10.

Eze, J. M., & Gill, L. S. (1992). Chromolaena odorata-a problematic weed. *Compositae Newsletter, 20*, 14–18.

Funk, V. A., Susanna, A., Steussy, T. F., & Robinson, H. E. (2009). Classification of compositae. In *Systematics, evolution, and biogeography of compositae* (pp. 171–189). Bratislava, Slovakia: International Association for Plant Taxonomy.

Gantayet, P. K., Adhikary, S. P., Lenka, K. C., & Padhy, B. (2014). Allelopathic impact of Lantana camara on vegetative growth and yield components of green gram (*Phaseolus radiatus*). *International Journal of Current Microbiology and Applied Sciences, 3*(7), 327–335.

Hernández Aro, M., Hernández, P. R., & Guillén, S. D. (2015). New micro bioassay sandwich to detection allelopathic activity from *Ipomoea batatas* (L.) Lam. *Journal of Food, Agriculture & Environment, 13*(3&4), 45–48.

Hossain, M. K., & Alam, M. N. (2010). Allelopathic effects of *Lantana camara* leaf extraction germination and growth behavior of some agricultural and forest crops in Bangladesh. *Pakistan Journal Weed Science and Research, 16*, 217–226.

Jain, A., Joshi, A., & Joshi, N. (2017). Allelopathic potential and HPTLC analysis of Ipomoea carnea. *International Journal of Life Science Scientific Research, 3*(5), 1278–1282.

Kamal, J. (2011). Impact of allelopathy of sunflower (*Helianthus annuus* L.) roots extract on physiology of wheat (*Triticum aestivum* L.). *African Journal of Biotechnology, 10*(65), 14465–14477.

Krause, M. S., Duarte, A. F., Merino, F. J., Paula, C. D., Miguel, M. D., & Miguel, O. G. (2016). Phytotoxic analysis of extract of leaves of *Solanum megalochiton* Mart. solanaceae on *Lactuca sativa* L. and *Allium cepa* L. *International Journal of Sciences, 5*(11), 36–42.

Mandal, M., Mamta, K., Mandal, S. S., Mandal, S. K., & Sinha, N. K. (2018). Allelopathic effect of aqueous leaf extracts of *Withania somnifera* dual on germination and seedling growth of wheat. *Journal of Pharmacognosy and Phytochemistry, 7*(1), 3158–3161.

Mawal, S. S., Shahnawaz, M., Sangale, M. K., & Ade, A. B. (2015). Assessment of allelopathic potential of the roots of *Parthenium hysterophorus* L. on some selected crops. *International Journal of Scientific Research in Knowledge, 3*(6), 145.

Mishra, A. (2015). Allelopathic properties of Lantana camara. *International Journal of Basic and Clinical Pharmacology, 3*, 13–28.

Mishra, S. K. (2018). Allelopathic potential of *Convolvulus arvensis* Linn. on seed germination and seedling growth of wheat (*Triticum vulgare*). *Research Journal of Pharmacognosy and Phytochemistry, 10*(2), 157–162.

Mushtaq, W., & Siddiqui, M. B. (2018). Allelopathy in Solanaceae plants. *Journal of Plant Protection Research, 58*(1), 1–7.

Netsere, A. (2015). Allelopathic effects of aqueous extracts of an invasive alien weed Parthenium hysterophorus L. on maize and sorghum seed germination and seedling growth. *Journal of Biology, Agriculture and Healthcare, 5*(1), 120–124.

Oudhia, P. (2000). Germination and seedling vigour of Kodomillet as affected by allelopathy of *Ipomoea carnea* Jacq. *Indian Journal of Plant Physiology, 5*(4), 383–384.

Oyeniyi, T. A., Odekanyin, O. O., Kuku, A., & Otusanya, O. O. (2016). Allelopathic effects of Tithonia diversifolia extracts on biochemical parameters and growth of *Vigna unguiculata*. *International Journal of Biology, 8*(3), 45.

Rajendiran, K., Yogeswari, D., & Arulmozhi, D. (2014). Allelopathic and cytotoxic effects of aqueous extracts of *Lantana camara* on *Vigna mungo* var. Vamban-16. *International Journal of Food, Agriculture and Veterinary Science, 4*(1), 179–185.

Ramachandran, A., & Venkataraman, N. S. (2016). Allelopathic effects of aqueous leaf extracts of Datura metel L. on Parthenium hysterophorus L. by A. Ramachandran and NS Venkataraman. *Life Sciences Leaflets, 72*, 14.

Ramona, Ş., Maria, I., Cărăbeţ, A., Ioana, G., Maria, V. A., & Manea, D. (2018). Allelopathic influence of *Datura stramonium* extracts on the germination and growing of soy plants. *Journal of Horticulture, Forestry and Biotechnology, 22*(2), 30–33.

Shikha, R., & Jha, A. K. (2016). Allelopathic effect of leaf extract of *Parthenium hysterophorus* L. on seed germination and growth of *Cicer aeritinum* L. *International Journal of Scientific Research, 5*(3), 652–655.

Shrivastav, S., & Jha, A. K. (2016). Effect of leaf extract of *Lantana camara* on growth of seedlings of *Cicer aeritinum*. *International Journal of Information Research and Review, 3*, 2612–2616.

Sinha, V. S. (2009). Nature. *Environment and Pollution Technology, 8*(4), 725–728.

Siyar, S., Muhammad, Z., Hussain, F., Hussain, Z., Islam, S., & Majeed, A. (2018). Allelopathic effects of two Asteraceae weeds (*Artemisia annua* and *Taraxacum officinalis*) on germination of maize and wheat. *PSM Biological Research, 23*(2), 44–47.

Sorecha, E. M., & Bayissa, B. (2017). Allelopathic effect of Parthenium hysterophorus L. on germination and growth of peanut and soybean in Ethiopia. *Advances in Crop Science and Technology, 5*(3), 1–4.

Srivastava, D., Shahi, S., & Kumar, R. (2015). Allelopathic effects of Ipomoea cairica (l) on noxious weed Parthenium hysterophorus Linn. *International Journal of Current Research in Life Sciences, 4*(12), 489–491.

Srivastava, D., & Shukla, K. (2016). Effect of leaves extract of *Ipomoea Cairica* on chlorophyll and carotenoid in *Parthenium hysterophorous* L. *International Journal of Research - GRANTHAALAYAH, 4*(4), 103–107.

Sunar, S., & Agar, G. (2017). Allelopathic effect of *Convolvulus arvensis* L. extracts on the phytohormones and cytological processes of *Zea mays* L. seeds. *European Journal of Experimental Biology, 7*, 15.

Teixeira de Oliveira, G., Amado, P. A., Siqueira Ferreira, J. M., & Alves Rodrigues dos Santos Lima L. (2019). Allelopathic effect of the ethanol extract and fractions of the aerial parts of *Lippia alba* (Verbenaceae). *Natural Product Research, 33*(16), 2402–2407.

Yuan, B., Li, S., Xiong, T., & Zhang, T. (2019). Cytogenetic and genotoxic effects of Ipomoea cairica (L.) sweet leaf aqueous extract on root growth of *Allium cepa* var. agrogarum (L.). *Allelopathy Journal, 46*(2), 205–214.

Chapter 5
Role of Allelochemicals in Agroecosystems

The biological invasion may discover its root into the marvels of allelopathy, the allelochemicals produced by the plants as a mean of their defense mechanism has turned out to be hindering to the growth of its native plants yet not to their coexisting species due to their adaptive capability (Callaway & Ridenour, 2004). In fact, allelopathy has been proposed as an outcome of long-term coevolution within established plant communities and may demonstrate considerable harmful impact on the newly introduced species (Mallik & Pellissier, 2000). Plant secondary metabolites can influence the ecological processes and structures and combat against the natural foes and competing plants, allelopathy may serve as a valuable strategy (Prince & Pohnert, 2010).

Allelopathy should also be considered as an important lead in agricultural ecosystems when dealing with the collective or successive cultivation of a variety of plants (Scognamiglio et al., 2013). However, allelochemicals have gained more popularity as a substitute for natural herbicides. Owing to its natural origin researchers proposed its incentive in the face of the allelochemical compound being more biodegradable and less polluting contrasting to the traditional herbicides, and inkling towards the suppression of weeds resistant to earlier used synthetic compounds (Reigosa, Pedrol, & González, 2006). Management of weeds through the utilization of allelopathy appeared in reference to its remarkable contribution to crop/weed interference (Belz, 2007; Farooq, Jabran, Cheema, Wahid, & Siddique, 2011). Practices like crop rotation and intercropping gained success in the light of allelochemicals, whereas the use of cover or smother crops is attributable to the weed suppression ability of allelopathic species. Reports suggest the inclusion of allelochemicals released by plants (Weston & Duke, 2003) or allelopathic crop water extracts (Farooq et al., 2011) in the field of weed management. Blends of organic natural products available commercially include vinegar, clove oil, and lemon extract mixtures. In addition, maize gluten meal is also accounted to serve as herbicide and fertilizer, ascribing to the herbicidal effect of its phytotoxic diterpenes and a pentapeptide (Liu & Christians, 1996). Further, the natural origin of the plants

W. Mushtaq et al., *Allelopathy*, SpringerBriefs in Agriculture,
https://doi.org/10.1007/978-3-030-40807-7_5

should not be confused with non-toxicity as the natural products can anyway be toxic as presented by fumonisin toxicity towards mammalian cells (Duke, Dayan, Romagni, & Rimando, 2000), and sorgoleone inciting dermatitis (Inderjit & Bhowmik, 2002).

Allelochemicals or allelochemics are known to mediate the phenomenon of allelopathy (Whittaker & Feeny, 1971). The term allelochemics was coined by Whittaker and Feeny (1971); secondary metabolites are also known to serve the purpose as these chemicals. The growth pattern of the donor species is variably regulated by these chemicals; the inhibitory effect at one concentration could be stimulatory for the same or distinct species at different concentrations (Narwal, 1994). Allelopathy is known to be a relatively new branch of science (Lal & Oudhia, 1999). Diverse types of interactions like crop–crop, crop–weed, weed–crop, and weed–weed can be readily understood with the help of allelopathy. Variability in the activity and concentration of the allelochemicals and its derivatives are known to occur over the growing season or with various parts of the plants viz. leaves, leaf mulch, leaf litter, flowers, stems, roots, barks, soil, soil leachates (Qasem & Foy, 2001; Macias, Molinillo, Varela, & Galindo, 2007; Jilani, Mahmood, Chaudhry, Hassan, & Akram, 2008; Gatti, Ferreira, Arduin, & Perez, 2010). Volatilization (Bertin, Yang, & Weston, 2003) and leaching (Bertin et al., 2003) serve the two means for the release of allelochemicals into the atmosphere or rhizosphere. Different means for the release of allelochemicals into the atmosphere or rhizosphere as per the favourable conditions are listed below:

1. Leaching (Bertin et al., 2003),
2. Volatilization (Bertin et al., 2003),
3. Decomposition of residues (Kohli, Singh, & Batish, 2001),
4. Root exudation (Bertin et al., 2003),
5. Pollen of some crop plants (Cruz-Ortega, Anaya, Hernández-Bautista, & Laguna-Hernández, 1998) and
6. Exposure to stress conditions, extreme temperature, drought, and UV exposure (Rice, 1984).

Allelochemicals based on their structural differences and properties (Li, Wang, Ruan, Pan, & Jiang, 2010) can be classified into the following categories (1) water-soluble organic acids, aliphatic aldehydes, straight-chain alcohols, and ketones; (2) polyacetylenes and long-chain fatty acids; (3) simple unsaturated lactones; (4) quinines (benzoquinone, complex quinines, and anthraquinone); (5) cinnamic acid and its derivatives; (6) coumarins; (7) phenolics; (8) tannins; (9) flavonoids; (10) terpenoids (sesquiterpene lactones, triterpenoids, and diterpenes) and steroids. Environmental conditions play a vital role in the variable quantitative production of allelochemicals (Bezuidenhout, 2012), some of which are mentioned as follows:

1. Light: The amount, duration, and intensity of light influence some allelochemicals. Under-storey plants tend to produce fewer allelochemicals in contrast with over-storey plants as the plants at greater heights filter out the ultraviolets whose exposure help in the production of the greatest quantities of these chemicals. Another factor that equally contributes to the most abundant

allelochemicals production is exposure to long-day photoperiods. Growing seasons also influence the amount of chemicals produced. The peak plant-growth period contributes to greater production than earlier or later growth season (Al-Jobori & Ali, 2014).

2. Mineral deficiency: Deficiency of minerals leads to more production of allelochemicals.
3. Drought stress: Under these conditions, the production of allelochemicals is enhanced.
4. Temperature: Cooler temperatures favour production in larger amounts. The site of production in the plant and its apparent effects in specific allelochemicals appear to be variable (Ali, 2008).
5. The age and type of plant tissue: Production of allelochemicals differs between species and within species as well. Therefore, it gains its importance during the process of extraction as a result of the irregular distribution of compounds in plants.
6. Plant diseases, shade and herbicides can also influence allelopathy.

Allelochemicals or phytotoxins are becoming a centre of interest to replace the existing weed management technologies as a result of their cheaper and more environmentally friendly attributes.

Thus, allelopathy has found itself as a new branch of herbicide development (Macias et al., 2007), as these compounds are either directly used as an herbicide or may present lead structures for herbicidal discovery (Duke et al., 2000).

Allelochemicals are considered as a suitable candidate as herbicides since they bear striking features of selectivity towards a lot of species especially towards weeds (Ratnadass, Fernandes, Avelino, & Habib, 2012). Among the different allelochemicals isolated from the plants, chemicals bearing the activity of suppression or elimination of competing plants received the most attention (Dayan, Cantrell, & Duke, 2009) and secondary metabolites having various molecular structures with phytotoxic activity have been isolated and also characterized from various sources (Scognamiglio, Esposito, et al., 2012, Scognamiglio, Fiumano, et al., 2012). Allelochemicals upon entering the cells undergo the process of retention, transportation, and/or transformation (Cheng, 1995). For the toxicity of inactive compounds of donor plants in its immediate vicinity, degradation and transformation play an important role (Rice, 1984). Apart from the degradation and transformation of chemical compounds present in the soil, various abiotic and biotic factors also make their contribution. Abiotic factors mainly include the physical and chemical factors such as light, heat, soil texture, soil organic and inorganic component (Einhellig, 1995), and microbes, viz. bacteria and fungi constrained for degradation/transport of organic molecules essentially in soil form the biotic components (Rice, 1984). Different allelochemicals have been distinguished from various weeds by many specialists: Scopoletin, p-hydroxybenzoic acid, coumarin, and vanillic acid are identified by Perez and Ormeno-Nunez (1991) as major allelochemicals of Avena fatua L. Dung, Nam, Huong, and Leclercq (1992) identified a-thujone from Artemisia vulgaris L. as a major allelochemical. Emodin and physcion from Polygonum sachalinense F. Schmidt are recognized by Inoue, Nishimura, Li, and Mizutani (1992). Table 5.1 depicts some of the identified allelochemicals from weeds (Reports from 2009).

Table 5.1 List of different plants with different allelochemicals identified (Reports from 2009)

Source weed	Chemical nature	Chemical name	Reference
Angelica sinensis (Oliv.) Diels	Monoterpenoid	4,6-dimethylbicyclo[3.1.0] hex-2-ene-2,6-dimethanol	Zhu, Wu, Wu, and Peng (2013)
Artemisia sieberi Besser., *Artemisia judaica* L., and *Origanum dayi* L.	Volatile allelochemicals	1,8-cineole, camphor, borneol germacrene D, artemisia alcohol, trans-thujone, para-cymene, samphene, sabinene, α-pinene, pinocarvone, benzoic acid (methyl vanillate), cis-sabinene hydrate, trans-sabinene hydrate, myrtenol, carvacrol, jasmine ketolactone, (Z)-methyl jasmonate, α-thujene, eugenol, cis-thujone, terpinen-4-ol, artemisia ketone, (E) ethyl cinnamate, davanone, artemisia alcohol, filifolide A, (Z) ethyl cinnamate, piperitone, β-davanone-2-ol, chrysanthenone, nordavanone, borneol, yomogi alcohol, camphor, methyl epi-jasmonate	Friedjung, Choudhary, Dudai, and Rachmilevitch (2013)
Avena fatua L.	Phenolic acids	p-coumaric acid, syringaldehyde, and vanillin	Fragasso, Platani, Miullo, Papa, and Iannucci (2012)
Avena fatua L.	Phenolic acids	Syringic acid, vanillin, 4-hydroxybenzoic acid, syringaldehyde, ferulic acid, p-coumaric acid and vanillic acid	Iannucci, Fragasso, Platani, and Papa (2013)
Avena fatua L.	Phenolic acids	Syringic acid, tricin, acacetin, syringoside, and diosmetin	Liu et al. (2016)
Bletilla striata (Thunb.) Rchb. F.	Glycosidic compounds	Militarine and dactylorhin A	Sakuno et al. (2010)
Cachrys pungens Jan	Flavonoids and phenolic acids	Naringin, quercetin, catechin, caffeic acid, ferulic acid, and gallic acid	Araniti et al. (2014)
Calamintha nepeta L. (Savi)	Phenolic acids	Gallic, vanillic, syringic, p-coumaric, and ferulic acids	Araniti, Lupini, Mercati, Statti, and Abenavoli (2013)
Carduus nutans L. and *Carduus acanthoides* L.	Allelochemical	Aplotaxene	Silva et al. (2014)
Castanea crenata Sieb. et Zucc	Allelochemical	2α,3β,7β,23-tetrahydroxyurs-12-ene-28-oic acid	Tuyen et al. (2018)
Centaurea diffusa Lam.	Sesquiterpene, fatty acid derived compounds	Caryophyllene oxide, linoleic acid, germacrene B, and aplotaxene	Quintana, El Kassis, Stermitz, and Vivanco (2009)

(continued)

Table 5.1 (continued)

Source weed	Chemical nature	Chemical name	Reference
Centaurea diffuse Lam	Quinoline	8-Hydroxyquinoline	Inderjit, Bajpai, and Rajeswari (2010)
Chenopodium murale L.	Phenolic acids	Vanillic, p-hydroxybenzoic, cinnamic, caffeic, protocatechuic, ferulic, and p-coumaric acids	Ghareib, Hamed, and Ibrahim (2010)
Cleome arabica L.	Sterol, flavonol, and dammarane type triterpene	β-Sitosterol, 17-α hydroxycabraleactone, amblyone, calycopterin, and 11-α-acetylbrachycarpone-22(23)-ene	Ladhari, Omezzine, Dellagreca, Zarrelli, and Haouala (2013)
Cleome viscosa L.	Lactam	Nonanoic acid LNA ((2-amino-9-(4-oxoazetidin-2-yl)-nonanoic acid)	Jana and Biswas (2011)
Diplostephium foliosissimum S.F. Blake	Hydroxycoumarin	Umbelliferone	Morikawa et al. (2011)
Echinochloa crus-galli (L.) P. Beauv	Phenolic acid	p-hydroxybenzoic acid	Esmaeili, Heidarzade, and Esmaeili (2012)
Euphorbia esula L.	Flavonoid	Tricin (5,7,4′-trihydroxy3′,5′-dimethoxyflavone)	Gomaa and Abdelgawad (2012)
Jasminum officinale L. f. var. grandiflorum L.	Secoiridoid glucoside	Oleuropein	Teerarak, Laosinwattana, and Charoenying (2010)
Lolium perenne L., *Dactylis glomerata* L., and *Rumex acetosa* L.	Benzoxazolinones and phenolic acid	Benzoxazolin-2(3H)-one (BOA) and cinnamic acid (CA)	Hussain, Gonzalez, and Reigosa (2011)

References

Ali, S. A (2008). *Effect of Dodonaea viscosa Jacq. residues on growth and yield of mungbean (Vigna mungo L. Hepper), maize (Zea mays L.) and their associated grassy weeds* (Ph.D. thesis, pp. 1–77). University of Baghdad, Iraq.

Al-Jobori, K. M., & Ali, S. A. (2014). Effect of Dodonaea viscosa Jacq. residues on growth and yield of mungbean (*Vigna mungo* L. Hepper). *African Journal of Biotechnology, 13*(24), 2407–2413.

Araniti, F., Lupini, A., Mercati, F., Statti, G. A., & Abenavoli, M. R. (2013). *Calamintha nepeta* L. (Savi) as source of phytotoxic compounds: Bio-guided fractionation in identifying biological active molecules. *Acta Physiologiae Plantarum, 35*(6), 1979–1988.

Araniti, F., Marrelli, M., Lupini, A., Mercati, F., Statti, G. A., & Abenavoli, M. R. (2014). Phytotoxic activity of *Cachrys pungens* Jan, a Mediterranean species: Separation, identification and quantification of potential allelochemicals. *Acta Physiologiae Plantarum, 36*(5), 1071–1083.

Belz, R. G. (2007). Allelopathy in crop/weed interactions—An update. *Pest Management Science, 63*(4), 308–326.

Bertin, C., Yang, X., & Weston, L. A. (2003). The role of root exudates and allelochemicals in the rhizosphere. *Plant and Soil, 256*(1), 67–83.

Bezuidenhout, S. R. (2012). *Allelopathy as a possible cause for crop yield reductions* (pp. 1–10). The department of agriculture and environmental affairs (DAEA).

Callaway, R. M., & Ridenour, W. M. (2004). Novel weapons: Invasive success and the evolution of increased competitive ability. *Frontiers in Ecology and the Environment, 2*(8), 436–443.

Cheng, H. H. (1995). Characterization of the mechanisms of allelopathy: Modeling and experimental approaches. In *Allelopathy: Organisms, processes and applications* (pp. 132–141). Washington, DC: American Chemical Society.

Cruz-Ortega, R., Anaya, A. L., Hernández-Bautista, B. E., & Laguna-Hernández, G. (1998). Effects of allelochemical stress produced by *Sicyos deppei* on seedling root ultrastructure of *Phaseolus vulgaris* and *Cucurbita ficifolia*. *Journal of Chemical Ecology, 24*(12), 2039–2057.

Dayan, F. E., Cantrell, C. L., & Duke, S. O. (2009). Natural products in crop protection. *Bioorganic and Medicinal Chemistry, 17*(12), 4022–4034.

Duke, S. O., Dayan, F. E., Romagni, J. G., & Rimando, A. M. (2000). Natural products as sources of herbicides: Current status and future trends. *Weed Research, 40*(1), 99–111.

Dung, N. X., Nam, V. V., Huong, H. T., & Leclercq, P. A. (1992). Chemical composition of the essential oil of *Artemisia vulgaris* L. var. indica maxim. From Vietnam. *Journal of Essential Oil Research, 4*(4), 433–434.

Einhellig, F. A. (1995). Allelopathy: Current status and future goals. In *Allelopathy: Organisms, process and application. American Chemical Society symposium series 582* (pp. 1–24). Washington, DC: American Chemical Society.

Esmaeili, M., Heidarzade, A., & Esmaeili, F. (2012). Quantifying of common allelochemicals in root exudates of barnyardgrass (*Echinochloa crus-galli* L.) and inhibitory potential against rice (*Oryza sativa*) cultivars. *American-Eurasian Journal of Agricultural and Environmental Sciences, 12*(6), 700–705.

Farooq, M., Jabran, K., Cheema, Z. A., Wahid, A., & Siddique, K. H. (2011). The role of allelopathy in agricultural pest management. *Pest Management Science, 67*(5), 493–506.

Fragasso, M., Platani, C., Miullo, V., Papa, R., & Iannucci, A. (2012). A bioassay to evaluate plant responses to the allelopathic potential of rhizosphere soil of wild oat (*Avena fatua* L.). *Agrochimica, 56*(2), 120–128.

Friedjung, A. Y., Choudhary, S. P., Dudai, N., & Rachmilevitch, S. (2013). Physiological conjunction of allelochemicals and desert plants. *PLoS One, 8*(12), 81580.

Gatti, A. B., Ferreira, A. G., Arduin, M., & Perez, S. C. G. D. A. (2010). Allelopathic effects of aqueous extracts of *Aristolochia esperanzae* O. Kuntze on development of *Sesamum indicum* L. seedlings. *Acta Botanica Brasilica, 24*(2), 454–461.

Ghareib, H. R. A., Hamed, M. S. A., & Ibrahim, O. H. (2010). Antioxidative effects of acetone fraction and vanillic acid from *Chenopodium murale* on tomato plants. *Weed Biology and Management, 10*, 64–72.

Gomaa, N. H., & Abdelgawad, H. R. (2012). Phytotoxic effects of *Echinochloa colona* (L.) link. (Poaceae) extracts on the germination and seedling growth of weeds. *Spanish Journal of Agricultural Research, 10*(2), 492–501.

Hussain, M. I., Gonzalez, L., & Reigosa, M. J. (2011). Allelopathic potential of Acacia melanoxylon on the germination and root growth of native species. *Weed Biology and Management, 11*(1), 18–28.

Iannucci, A., Fragasso, M., Platani, C., & Papa, R. (2013). Plant growth and phenolic compounds in the rhizosphere soil of wild oat (*Avena fatua* L.). *Frontiers in Plant Science, 4*, 1–7.

Inderjit, Bajpai, D., & Rajeswari, M. S. (2010). Interaction of 8-hydroxyquinoline with soil environment mediates its ecological function. *Plos One, 5*(9), 1–7.

Inderjit, & Bhowmik, P. C. (2002). The importance of allelochemicals in weed invasiveness and the natural suppression. In *Chemical ecology of plant: Allelopathy of aquatic and terrestrial ecosystems* (pp. 187–192). Basel, Switzerland: Birkhauser Verlag.

Inoue, M., Nishimura, H., Li, H. H., & Mizutani, J. (1992). Allelochemicals from *Polygonum sachalinense* Fr. Schm. (Polygonaceae). *Journal of Chemical Ecology, 18*(10), 1833–1840.

Jana, A., & Biswas, S. M. (2011). Lactam nonanic acid, a new substance from *Cleome viscosa* with allelopathic and antimicrobial properties. *Journal of Biosciences, 36*(1), 27–35.

Jilani, G., Mahmood, S., Chaudhry, A. N., Hassan, I., & Akram, M. (2008). Allelochemicals: Sources, toxicity and microbial transformation in soil—A review. *Annals of Microbiology, 58*(3), 351–357.

Kohli, R. K., Singh, H. P., & Batish, D. R. (2001). *Allelopathy in agroecosystems* (Vol. 4, p. 447). New York, NY: Food Products Press.

Ladhari, A., Omezzine, F., Dellagreca, M., Zarrelli, A., & Haouala, R. (2013). Phytotoxic activity of *Capparis spinosa* L. and its discovered active compounds. *Allelopathy Journal, 32*(2), 175–190.

Lal, B., & Oudhia, P. (1999). Beneficial effects of allelopathy. I. crop production. *Indian Journal of Weed Science, 31*(1–2), 103–105.

Li, Z. H., Wang, Q., Ruan, X., Pan, C. D., & Jiang, D. A. (2010). Phenolics and plant allelopathy. *Molecules, 15*(12), 8933–8952.

Liu, D. L., & Christians, N. E. (1996). Bioactivity of a pentapeptide isolated from corn gluten hydrolysate on *Lolium perenne* L. *Journal of Plant Growth Regulation, 15*(1), 13–17.

Liu, X., Tian, F., Tian, Y., Wu, Y., Dong, F., Xu, J., & Zheng, Y. (2016). Isolation and identification of potential allelochemicals from aerial parts of *Avena fatua* L. and their allelopathic effect on wheat. *Journal of Agricultural and Food Chemistry, 64*(18), 3492–3500.

Macias, F. A., Molinillo, J. M., Varela, R. M., & Galindo, J. C. (2007). Allelopathy-a natural alternative for weed control. *Pest Management Science, 63*(4), 327–348.

Mallik, A. U., & Pellissier, F. (2000). Effects of Vaccinium myrtillus on spruce regeneration: Testing the notion of coevolutionary significance of allelopathy. *Journal of Chemical Ecology, 26*(9), 2197–2209.

Morikawa, C. I. O., Miyaura, R., Kamo, T., Hiradate, S., Perez, J. A. C., & Fujii, Y. (2011). Isolation of umbelliferone as a principal allelochemical from the Peruvian medicinal plant *Diplostephium foliosissimum* (Asteraceae). *Revista de la Sociedad Quimica del Peru, 77*(4), 285–291.

Narwal, S. S. (1994). *Allelopathy in crop production* (p. 288). Jodhpur, Rajasthan: Scientific Publishers.

Perez, F. J., & Ormeno-Nunez, J. (1991). Root exudates of wild oats: Allelopathic effect on spring wheat. *Phytochemistry, 30*(7), 2199–2202.

Prince, E. K., & Pohnert, G. (2010). Searching for signals in the noise: Metabolomics in chemical ecology. *Analytical and Bioanalytical Chemistry, 396*(1), 193–197.

Qasem, J. R., & Foy, C. L. (2001). Weed allelopathy, its ecological impacts and future prospects: A review. *Journal of Crop Production, 4*(2), 43–119.

Quintana, N., El Kassis, E. G., Stermitz, F. R., & Vivanco, J. M. (2009). Phytotoxic compounds from roots of *Centaurea diffusa* Lam. *Plant Signaling and Behavior, 4*(1), 9–14.

Ratnadass, A., Fernandes, P., Avelino, J., & Habib, R. (2012). Plant species diversity for sustainable management of crop pests and diseases in agroecosystems: A review. *Agronomy for Sustainable Development, 32*(1), 273–303.

Reigosa, M. J., Pedrol, N., & González, L. (2006). *Allelopathy: A physiological process with ecological implications* (pp. 1–637). Dordrecht, The Netherlands: Springer.

Rice, E. L. (1984). *Allelopathy* (2nd ed., p. 421). New York, NY: Academic Press.

Sakuno, E., Kamo, T., Takemura, T., Sugie, H., Hiradate, S., & Fujii, Y. (2010). Contribution of militarine and dactylorhin a to the plant growth-inhibitory activity of a weed-suppressing orchid, *Bletilla striata*. *Weed Biology and Management, 10*(3), 202–207.

Scognamiglio, M., D'Abrosca, B., Esposito, A., Pacifico, S., Monaco, P., & Fiorentino, A. (2013). Plant growth inhibitors: Allelopathic role or phytotoxic effects? Focus on Mediterranean biomes. *Phytochemistry Reviews, 12*(4), 803–830.

Scognamiglio, M., Esposito, A., D'Abrosca, B., Pacifico, S., Fiumano, V., Tsafantakis, N., & Fiorentino, A. (2012). Isolation, distribution and allelopathic effect of caffeic acid derivatives from *Bellis perennis* L. *Biochemical Systematics and Ecology, 43*, 108–113.

Scognamiglio, M., Fiumano, V., D'Abrosca, B., Pacifico, S., Messere, A., Esposito, A., & Fiorentino, A. (2012). Allelopathic potential of alkylphenols from *Dactylis glomerata* subsp. hispanica (Roth) Nyman. *Phytochemistry Reviews, 5*(1), 206–210.

Silva, F. M., Donega, M. A., Cerdeira, A. L., Corniani, N., Velini, E. D., Cantrell, C. L., & Duke, S. O. (2014). Roots of the invasive species *Carduus nutans* L. and C. *acanthoides* L. produce large amounts of aplotaxene, a possible allelochemical. *Journal of Chemical Ecology, 40*(3), 276–284.

Teerarak, M., Laosinwattana, C., & Charoenying, P. (2010). Evaluation of allelopathic, decomposition and cytogenetic activities of *Jasminum officinale* L. f. var. grandiflorum (L.) Kob. on bioassay plants. *Bioresource Technology, 101*(14), 5677–5684.

Tuyen, P. T., Xuan, T. D., Tu-Anh, T. T., Mai, V. T., Ahmad, A., Elzaawely, A. A., & Khanh, T. D. (2018). Weed suppressing potential and isolation of potent plant growth inhibitors from *Castanea crenata* Sieb. et Zucc. *Molecules, 23*(2), 345.

Weston, L. A., & Duke, S. O. (2003). Weed and crop allelopathy. *Plant Science, 22*, 367–389.

Whittaker, R. H., & Feeny, P. P. (1971). Allelochemics: Chemical interactions between species. *Science, 171*(3973), 757–770.

Zhu, H., Wu, S., Wu, Q., & Peng, C. (2013). Isolation and identification of autotoxic chemicals from *Angelica sinensis* (Oliv.) diels. *Journal of Food, Agriculture and Environment, 11*(3 and 4), 2136–2140.

Chapter 6
Allelopathic Control of Native Weeds

Weeds are undesired plants that are of no economical use and are hard to manage by farmers. Weeds affect the growth and development of crops and therefore limit their productivity (Ani, Onu, Okoro, & Uguru, 2018). In the agricultural system, weeds compete with crop plants resulting in the loss of their yield (Gaba, Reboud, & Fried, 2016). They limit the accessibility of light, moisture, space to crops and deteriorate their quality (Guglielmini, Verdú, & Satorre, 2017). In view of these features, it has become necessary to check its growth. However, with the beginning of agriculture, the most prominent weed control approaches include an application of herbicides and hand/motorized weeding (Jabran, Mahajan, Sardana, & Chauhan, 2015; Young, Meyer, & Woldt, 2014). These approaches have a remarkable contribution to the improvement of crop production; but various hurdles are associated with them, as well. However, wide utilization of herbicides to check the growth of weeds has led to severe ecological and environmental problems like herbicide resistance, a shift in weed flora, and environmental pollution and health hazards due to their toxic residues in soil, water, and food chain. The harmful effect of commercial herbicides makes it suitable to explore various other weed management alternatives (Nirmal Kumar, Amb, & Bora, 2010) and allelopathy seems to be one of the options (Rawat, Maikhuri, Bahuguna, Jha, & Phondani, 2017). Allelopathy is an eco-friendly weed management tool, which is practiced to combat the impacts of environmental pollution. Allelopathy is a chemical method that allows the plant to compete for a narrow range of resources (Gioria & Osborne, 2014).

Allelopathy is an ecological process in which a plant through the release of chemicals, impact growth behaviour, physiology, and development of other plants which are living in its proximity (Hussain, Ilahi, Malik, Dasti, & Ahmad, 2011). Allelopathy has deep significance in inhibiting growth weed and is among the productive methods to control weeds (Zeng, 2014). Allelochemicals are the active media of allelopathy which are non-nutritive, a secondary metabolite of plants formed as byproducts during distinct physiological processes in plants (Bhadoria, 2011).

W. Mushtaq et al., *Allelopathy*, SpringerBriefs in Agriculture,
https://doi.org/10.1007/978-3-030-40807-7_6

6.1   Allelopathic Crops

Various crop species have been reported to possess the potential of allelopathy and are used to control the intensity of weeds in fields (Jabran & Farooq, 2013). Several crops with strong allelopathic traits have been screened by researchers. Some of the common field crops and weeds suppressed by these crops are given in the following table;

Crops	Weeds suppressed	References
Oryza sativa L	*Echinochloa crus-galli* P. Beauv	Khanh et al. (2017)
Tagetes minuta	*Amaranthus tricolor,* *Echinochloa crus-galli*	Arora, Batish, Kohli, and Singh (2017)
Canola	*Lolium rigidum* Gaudin	Asaduzzaman, An, Pratley, Luckett, and Lemerle (2014)
Wheat	*Descurainia sophia* (L.) Webb ex Prantl	Zuo, Li, Ma, and Yang (2014)
Fennel, rue, and sage	*Lepidium draba*	Ravlić, Baličević, Nikolić, and Sarajlić (2016)
Sinapis alba and *Brassica rapa*	*Avena fatua*	El-Rokiek Kowthar, Ahmed, Messiha, Mohamed, and El-Masry (2017)
Aloe vera	*Tripleurospermum inodorum,* *Amaranthus retroflexus*	Ravlić, Baličević, Visković, and Smolčić (2017)
Artemisia absinthium, *Psidium guajava*	*Parthenium hysterophorus*	Kapoor et al. (2019)
Petroselinum crispum *(Mill.)*	*Ambrosia artemisiifolia* L, *Polygonum aviculare* L	Rodino, Buțu, and Buțu (2016)
Rye	*Chenopodium album, Abutilon theophrasti* Medik.	Bernstein, Stoltenberg, Posner, and Hedtcke (2014)
White mustard	*Amaranthus blitoides* S.Watson, C. album	Alcantara, Pujadas, and Saavedra (2011)

6.2   Allelopathic Compounds

A large number of allelopathic substances have been isolated and characterized by plants; however, only a small number of allelochemicals have been unveiled for their mode of action (Vyvyan, 2002). A large number of allelochemicals isolated from plants have been used for the discovery of herbicide and have produced successful compounds. The triketone herbicides (mesotrione, tembotrione, sulcotrione) are chemical analogs of the allelochemicals leptospermone, grandiflorone, and flavesone. These compounds inhibit hydroxyphenylpyruvate dioxygenase (HPPD) enzyme (Dayan, Owen, & Duke, 2012). It has been reported that numerous allelochemicals such as glucosinolates and phenolic compounds (*Brassica* sp., black mustard), fatty acids (*Polygonum spp.*), isoflavonoids and phenolics (*Trifolium spp.*, *Melilotus spp.*), phenolic acids and scopoletin (*Avena sativa*), hydroxamic acids and benzoxazinoids (*Triticum* sp.), phenolic, acids, dhurrin, and sorgoleone (*Sorghum bicolor*) are utilized for weed control (Jabran, 2017).

6.3 Allelopathic Weed Control

Weed control can be achieved by growing the allelopathic plants or putting allelopathic material from plant residues close to weeds which leads to production of allelochemicals (Ferreira & Reinhardt, 2016). Allelochemicals released by the decomposing plant material are absorbed by the target weeds. Growing of allelopathic plants in a field for a specific time period until their roots exudate allelochemicals is also employed to control weed growth. The most common practice utilized for allelopathic weed control is crop rotation (Farooq, Jabran, Cheema, Wahid, & Siddique, 2011). Another method involves the dipping of allelopathic chaff in water for a specific period of time to get allelochemicals in a liquid-solution. This method of weed control has been recommended by various researchers and is utilized either alone or in combination with other methods of weed control (Khan, Ali, et al., 2012; Khan, Khan, et al., 2012). Many studies also specify the positive effects of allelopathy on the soil environment by enhancing nutrient availability to crop plants through increased activities of soil microbes (Wang, Li, Jhan, Weng, & Chou, 2013; Zeng, 2014).

6.4 Approaches for Allelopathic Weed Management

6.4.1 Intercropping

Intercropping is the practice of raising crops together at the same time in the same field. It is a key approach for enhancing input (land, fertilizer, and water) use efficiency and crop yield and economic returns (Khan, Ali, et al., 2012; Khan, Khan, et al., 2012). Moreover, intercropping with allelopathic crops can bestow eco-friendly substitute to chemical weed control (Jabran & Chauhan, 2018). Primarily, allelopathic crops intercropped with other crop plants assist in decreasing weed intensity, thereby increasing crop productivity. Several allelopathic crop species intercropped with maize checked the growth of different narrow- and broad-leaved weed species (Nawaz, Farooq, Cheema, & Cheema, 2014). Similarly, the intercropping of white clover (*Trifolium repens* L.), black medic (*Medicago lupulina* L.), alfalfa, and red clover (*Trifolium pratense* L.) in the wheat crop was productive in terms of weed control and also increased wheat yield (Amosse, Jeuffroy, Celette, &

David, 2013). Reduced-intensity of *Orobanche crenata* Forssk was reported by Fernández-Aparicio, Emeran, and Rubiales (2010) when berseem was raised along with legumes (broad bean and pea).

6.4.2 Cover Crops

Allelopathic cover crops, in addition to their various benefits like protecting plants from soil erosion, refining soil fertility and structure, nitrogen fixation, are also successfully utilized in weed control (Tursun, Işık, Demir, & Jabran, 2018). Due to their competitive and allelopathic effects, these crops are utilized for suppression of weeds in agricultural fields (Sturm, Peteinatos, & Gerhards, 2018). Some of the important cover crops include canola, rapeseed, cereal rye, crimson clover, wheat, red clover, brown mustard, oats, cowpea, fodder radish, annual ryegrass, mustards, buckwheat, hairy vetch, and black mustard. Several cropping systems (e.g., organic cropping) depend on cover cropping for weed management (Mirsky et al., 2013). It was reported that cover crop mixtures of *Sinapis alba* L., (Raphanus sativus var. niger J. Kern) and Vicia sativa suppressed weeds. These authors suggest that biochemical effects are responsible for weed suppression of cover crops. Cover crops smother weed growth either by forming a physical barrier decreasing the intensity of light and temperature or are responsible for the allelochemical release from residues and microbes. They outgrow the weeds and out shade them. It has been noted that allelopathic crop plants substantially decrease weed crops. Potential of cover crops for controlling weeds is dependent upon its allelopathic potential and duration in the field, allelopathic crop for long duration in the field allows effective weed control (Bhowmik, 2003).

6.4.3 Residue Incorporation

Allelopathic plant residues left in the field either unintentionally or added manually exhibit the potential of suppressing weed intensity. The decomposing plant residue releases a number of allelochemicals mostly phenolics that suppress weed activity. For example, barley, rye, and triticale crop residues kept in a maize field were reported for having an allelopathic effect on *E. crus-galli* and *Setaria verticillata* (L.) P. Beauv. in Greece (Dhima, Vasilakoglou, Eleftherohorinos, & Lithourgidis, 2006) where the allelopathic mulches reduced the growth of *S. verticillata* and *E. crus-galli* in comparison to non-mulched treatment. Moreover, applied mulches had no negative effect on maize. However, barley mulches applied to the maize field increased the grain yield by 45% in the plots compared with the non-treated control (Dhima et al., 2006). Bajgai, Kristiansen, Hulugalle, and McHenry (2015) reported a remarkable reduction of weed infestation in broccoli (Brassica oleracea L.) crop when maize residues were added in the field after its harvest. Similarly, weed control potential in the field crops has been reported with sunflower residues and surface mulches (Rawat et al., 2017).

References

Alcantara, C., Pujadas, A., & Saavedra, M. (2011). Management of *Sinapis alba* subsp. mairei winter cover crop residues for summer weed control in southern Spain. *Crop Protection, 30*, 1239–1244.

Amosse, C., Jeuffroy, M. H., Celette, F., & David, C. (2013). Relay-intercropped forage legumes help to control weeds in organic grain production. *European Journal of Agronomy, 49*, 158–167.

Ani, O., Onu, O., Okoro, G., & Uguru, M. (2018). Overview of biological methods of weed control. *Biological Approaches for Controlling Weeds, 5*, 5.

Arora, K. O., Batish, D. A., Kohli, R., & Singh, H. (2017). Allelopathic impact of essential oil of Tagetes minuta on common agricultural and wasteland weeds. *Innovare Journal of Agricultural Science, 5*, 1–4.

Asaduzzaman, M., An, M., Pratley, J. E., Luckett, D. J., & Lemerle, D. (2014). Canola (*Brassica napus*) germplasm shows variable allelopathic effects against annual ryegrass (*Lolium rigidum*). *Plant and Soil, 380*(1–2), 47–56.

Bajgai, Y., Kristiansen, P., Hulugalle, N., & McHenry, M. (2015). Comparison of organic and conventional managements on yields, nutrients and weeds in a corn–cabbage rotation. *Renewable Agricultural and Food Systems, 30*(2), 132–142.

Bernstein, E. R., Stoltenberg, D. E., Posner, J. L., & Hedtcke, J. L. (2014). Weed community dynamics and suppression in tilled and no-tillage transitional organic winter rye–soybean systems. *Weed science, 62*(1), 125–137.

Bhadoria, P. B. (2011). Allelopathy: A natural way towards weed management. *American Journal of Experimental Agriculture, 1*(1), 7.

Bhowmik, P. C. (2003). Challenges and opportunities in implementing allelopathy for natural weed management. *Crop protection, 22*(4), 661–671.

Dayan, F. E., Owen, D. K., & Duke, S. O. (2012). Rationale for a natural products approach to herbicide discovery. *Pest Management Science, 68*, 519–528.

Dhima, K. V., Vasilakoglou, I. B., Eleftherohorinos, I. G., & Lithourgidis, A. S. (2006). Allelopathic potential of winter cereals and their cover crop mulch effect on grass weed suppression and corn development. *Crop Science, 46*, 345–352.

El-Rokiek Kowthar, G., Ahmed, S. A., Messiha, N. K., Mohamed, S. A., & El-Masry, R. R. (2017). Controlling the grassy weed *Avena fatua* associating wheat plants with the seed powder of two brassicaceae plants *Brassica rapa* and *Sinapis alba*. *Middle East Journal, 6*(4), 1014–1020.

Farooq, M., Jabran, K., Cheema, Z. A., Wahid, A., & Siddique, K. H. M. (2011). The role of allelopathy in agricultural pest management. *Pest Management Science, 67*, 493–506.

Fernández-Aparicio, M., Emeran, A. A., & Rubiales, D. (2010). Inter-cropping with berseem clover (*Trifolium alexandrinum*) reduces infection by *Orobanche crenata* in legumes. *Crop Protection, 29*(8), 867–871.

Ferreira, M. I., & Reinhardt, C. F. (2016). Allelopathic weed suppression in agroecosystems: A review of theories and practices. *African Journal of Agricultural Research, 11*(6), 450–459.

Gaba, S., Reboud, X., & Fried, G. (2016). Agroecology and conservation of weed diversity in agricultural lands. *Botany Letters, 163*(4), 351–354.

Gioria, M., & Osborne, B. A. (2014). Resource competition in plant invasions: Emerging patterns and research needs. *Frontiers in Plant Science, 5*, 501.

Guglielmini, A. C., Verdú, A. M., & Satorre, E. H. (2017). Competitive ability of five common weed species in competition with soybean. *International journal of pest management, 63*(1), 30–36.

Hussain, F., Ilahi, I., Malik, S. A., Dasti, A. A., & Ahmad, B. (2011). Allelopathic effects of rain leachates and root exudates of *Cenchrus ciliaris* L. and *Bothriochloa pertusa* (L.) A. Camus. *Pakistan Journal of Botany, 43*(1), 341–350.

Jabran, K. (2017). *Manipulation of allelopathic crops for weed control* (1st ed.). Cham, Switzerland: Springer.

Jabran, K., & Chauhan, B. S. (2018). *Non-chemical weed control* (1st ed.). New York, NY: Elsevier.

Jabran, K., & Farooq, M. (2013). Implications of potential allelopathic crops in agricultural systems. In *Allelopathy* (pp. 349–385). Berlin, Germany: Springer.

Jabran, K., Mahajan, G., Sardana, V., & Chauhan, B. S. (2015). Allelopathy for weed control in agricultural systems. *Crop Protection, 72*, 57–65.

Kapoor, D., Tiwari, A., Sehgal, A., Landi, M., Brestic, M., & Sharma, A. (2019). Exploiting the allelopathic potential of aqueous leaf extracts of *Artemisia absinthium* and *Psidium guajava* against *Parthenium hysterophorus*, a widespread weed in India. *Plants, 8*(12), 552.

Khan, M. A., Ali, K., Hussain, Z., & Afridi, R. A. (2012). Impact of maize-legume intercropping on weeds and maize crop. *Pakistan Journal of Weed Science Research, 18*(1), 127–136.

Khan, M. B., Khan, M., Hussain, M., Farooq, M., Jabran, K., & Lee, D. J. (2012). Bio-economic assessment of different wheat-canola intercropping systems. *International Journal of Agriculture and Biology, 14*, 769–774.

Khanh, T. D., Son, D. B., Phuong, V. T., Hoa, L. T., Linh, L. H., Yen, N. V., & Trung, K. H. (2017). Assessment of weed-suppressing potential among rice (*Oryza sativa* L.) landraces against the growth of Barnyardgrass (*Echinochloa crus-galli* P. Beauv) in field condition. *Academy of Agriculture Journal, 2*(08), 47–51.

Mirsky, S. B., Ryan, M. R., Teasdale, J. R., Curran, W. S., Reberg-Horton, C. S., Spargo, J. T., … Moyer, J. W. (2013). Overcoming weed management challenges in cover crop–based organic rotational no-till soybean production in the eastern United States. *Weed Technology, 27*(1), 193–203.

Nawaz, A., Farooq, M., Cheema, S. A., & Cheema, Z. A. (2014). Role of allelopathy in weed management. In *Recent advances in weed management* (pp. 39–61). New York, NY: Springer.

Nirmal Kumar, J. I., Amb, M. K., & Bora, A. (2010). Chronic response of *Anabaena fertilissima* on growth, metabolites and enzymatic activities by chlorophenoxy herbicide. *Pesticide Biochemistry and Physiology, 98*(2), 168–174.

Ravlić, M., Baličević, R., Nikolić, M., & Sarajlić, A. (2016). Assessment of allelopathic potential of fennel, rue and sage on weed species hoary cress (*Lepidium draba*). *Notulae Botanicae Horti Agrobotanici Cluj-Napoca, 44*(1), 48–52.

Ravlić, M., Baličević, R., Visković, M., & Smolčić, I. (2017). Response of weed species on allelopathic potential of Aloe vera (L.) Burm. f. *Herbologia, 16*(2), 49–55.

Rawat, L. S., Maikhuri, R. K., Bahuguna, Y. M., Jha, N. K., & Phondani, P. C. (2017). Sunflower allelopathy for weed control in agriculture systems. *Journal of Crop Science and Biotechnology, 20*(1), 45–46.

Rodino, S., Buțu, M., & Buțu, A. (2016). Comparative study on allelopathic potential of Petroselinum crispum (MILL.) FUSS. *Analele Stiintifice ale Universitatii "Al. I. Cuza" din Iasi, 62*(2), 35.

Sturm, D. J., Peteinatos, G., & Gerhards, R. (2018). Contribution of allelopathic effects to the overall weed suppression by different cover crops. *Weed Research, 58*(5), 331–337.

Tursun, N., Işık, D., Demir, Z., & Jabran, K. (2018). Use of living, mowed, and soil-incorporated cover crops for weed control in apricot orchards. *Agronomy, 8*(8), 150.

Vyvyan, J. R. (2002). Allelochemicals as leads to herbicides and agrochemicals. *Tetrahedron, 58*, 1631–1646.

Wang, C. M., Li, T. C., Jhan, Y. L., Weng, J. H., & Chou, C. H. (2013). The impact of microbial biotransformation of catechin in enhancing the allelopathic effects of *Rhododendron formosanum*. *PLoS One, 8*(12), e85162.

Young, S. L., Meyer, G. E., & Woldt, W. E. (2014). Future directions for automated weed management in precision agriculture. In *Automation: The future of weed control in cropping systems* (pp. 249–259). Dordrecht, The Netherlands: Springer.

Zeng, R. S. (2014). Allelopathy-the solution is indirect. *Journal of Chemical Ecology, 40*(6), 515–516.

Zuo, S., Li, X., Ma, Y., & Yang, S. (2014). Soil microbes are linked to the allelopathic potential of different wheat genotypes. *Plant and Soil, 378*(1–2), 49–58.

Chapter 7
Mechanism of Action of Allelochemicals

Allelopathy could be characterized as "an imperative component of plant impedance interceded by the addition of secondary metabolites produced by plants into the soil rhizosphere" (Weston, 2005). These secondary metabolites are typically exuded into the rhizosphere and affect the development of plants that are growing in the vicinity of allelopathic plants (Akemo, Regnier, & Bennett, 2000). Chemical compounds that inflict allelopathic impacts are called allelochemicals or allelochemics, which are by and large considered to be those chemical groups, for example, alkaloids, flavonoids, glucosinolates, phenolics and terpenoids (Reigosa, Souto, & Gonz, 1999). Natural products recognized with allelopathic potential have been classified into the following groups: (a) cytotoxic gases, (b) organic acids, (c) aromatic acids, (d) simple unsaturated lactones, (e) coumarins, (f) quinones, (g) flavonoids, (h) tannins, (i) alkaloids, and (j) terpenoids and steroids (Mushtaq & Siddiqui, 2018). Allelopathy has been well documented for many great years (Rice, 1984), however, the understanding of the mechanisms of the mode of action of allelochemicals stays darken (Mohamadi & Rajaie, 2009). Several biosynthetic pathways are responsible for the production of the various classes of these chemical compounds, though they are not necessary for primary processes of growth and reproduction for the allelopathic species (Pagare, Bhatia, Tripathi, Pagare, & Bansal, 2015). However, these compounds can influence plant development indirectly by modifying the interspecific competition for the plants in association (Abhilasha, Quintana, Vivanco, & Joshi, 2008). A wide array of these compounds are known today, however, just a limited number has been recognized as allelochemicals (Mushtaq & Siddiqui, 2018; Rice, 1984). Allelochemicals are predominantly present throughout the plant including leaves, stems, roots, rhizomes, inflorescence, pollen, fruits and seeds (An, Pratley, & Haig, 1998). The production of allelochemicals in a plant species may vary spatially and over time scale; Singh, Jhaldiyal, and Kumar (2009) found foliar and leaf litter leachates of *Eucalyptus* species more lethal than its bark leachates to some commercial crops.

© The Author(s), under exclusive license to Springer Nature Switzerland AG 2020
W. Mushtaq et al., *Allelopathy*, SpringerBriefs in Agriculture,
https://doi.org/10.1007/978-3-030-40807-7_7

As indicated by Rice (1984) and Putnam (1985), there are four ways by which these biosynthetic substances are discharged: (a) Volatilization, discharge into the air. It is accountable only if it were dry or semi-arid conditions. The compounds might be absorbed as vapours by encompassing plants, be absorbed from the condensate of these vapours in dew or the condensate may reach the soil and absorbed by the roots. (b) Leaching, through precipitation, dew or irrigation which may ooze out the allelochemicals from the aboveground parts of plants and subsequently deposit them on other plants or into the soil. Leaching may likewise happen through plant residues. (c) Root exudation from plant roots contributes one of the major direct inputs into the rhizosphere soil. Regardless of whether these compounds are effectively oozed, spilled or arise from dead cells sloughing off the roots isn't comprehended however as of now and, (d) Decomposition of plant residues, it is hard to decide if these toxic substances are contained in residues and discharged upon decay, or microorganisms convert the simple residues into these toxic products as a result of the presence of microbial enzymes (Mushtaq & Siddiqui, 2018). With these couple of mechanisms introduced above, you can see the immense variations that exist for dispersing biosynthetic allelochemicals in different allelopathic species (Khalid, Ahmad, & Shad, 2002).

Allelopathic intrusions often are an outcome of the joint activity of different compounds. No single phytotoxin was exclusively in charge of or delivered allelopathy because of intervention by a neighboring plant (Weston, 2005). Allelochemicals found to hinder the development of a species at a certain concentration may enhance the development of the same species or another at a different concentration. The response of different species to different allelochemicals is concentration-dependent, the degree of inhibition increases with the increasing concentration (Gulzar, Siddiqui, & Bi, 2016; Ishak & Sahid, 2014; Mushtaq, Ain, & Siddiqui, 2018; Mushtaq, Ain, Siddiqui, & Hakeem, 2019). The selective action of tree allelochemicals on crops and different plants has been reported as well. Leachates of the elder tree can impede the development of pangola grass however enhance the growth of bluestem, another pasture grass (Singh, Uniyal, & Todaria, 2008). As indicated by Cheng (1992) once the allelochemicals are discharged by the donor plant into the environment, various associating processes will happen. These processes have been distinguished as; (a) Retention, the impeded movement of the substance through soil, water and air from one area to another; (b) Transformation, the alteration in form or structure of the compound, prompting to its fractional change or total decomposition and, (c) Transport, describes how the compounds move in the environment. Environmental factors, soil properties, nature of the compound and species involved in the interaction jointly influence these processes. Moreover, the fate of the allelochemicals depends on the interactions and kinetics of individual processes spatially, at a specific site under a specific set of natural conditions (Rizvi, 2012).

The mode of activity of a substance could comprehensively be partitioned into direct and indirect action (Blum, 2002). Influences through the modification of soil properties, nutritional status and an altered population or role of microorganisms and nematodes express indirect action. The direct action comprises the biochemical/

physiological influences of allelochemicals on different essential processes of plant metabolism. Procedures affected by allelochemicals include:

1. Mineral absorption: allelochemicals can modify the rate of ion absorption by plants (Baar, Ozinga, Sweers, & Kuyper, 1994). Phenolic acids reduce the uptake of both macro and micronutrients (Akemo et al., 2000).
2. Cytology and ultrastructure: an assortment of allelochemicals have been appeared to repress mitosis in plant roots (Celik & Aslanturk, 2010; Gulzar et al., 2016; Mohamadi & Rajaie, 2009; Mushtaq et al., 2019; Teerarak, Laosinwattana, & Charoenying, 2010).
3. Phytohormones: the plant growth hormones, IAA (Indole acetic acid) and GA (gibberellins) control cell enlargement in plants. IAA is available in both active and inactive forms (inactivated by IAA-oxidase). IAA-oxidase is repressed by different allelochemicals (Chou, 1980). Ethylene and ABA (abscisic acid) production increased upon allelopathy stress (Bogatek, Oracz, & Gniazdowska, 2005)
4. Membrane permeability: Various organic compounds exert their influence through the modification of membrane permeability (Galindo et al., 1999). Exudation of chemicals from roots on root cuts has been utilized as a penetrability index since plant membranes are hard to study (Gniazdowska & Bogatek, 2005).
5. Photosynthesis: benzoic and cinnamic acid reduced chlorophyll content in soyabean thus inhibited photosynthesis (Baziramakenga, Simard, & Leroux, 1994). Photosynthetic inhibitors might be electron inhibitors or uncouplers, energy-exchange inhibitors electron acceptors or a mixture of the above (Batish, Singh, & Kaur, 2001)
6. Respiration: allelochemicals can fortify or repress respiration, both of which can be injurious to the vitality of the energy-producing system (Batish et al., 2001)
7. Protein synthesis: Studies using radio-marked C^{14} sugars or amino acids, and traced their incorporation into protein, reported that allelochemicals obstruct protein synthesis (Bertin et al., 2007).
8. Enzyme activity: Rice (1984) gave an account of various allelochemicals that restrain the specific enzyme activity in the plants (Muscolo, Panuccio, & Sidari, 2001). *N. plumbaginifolia* allelochemicals stimulated the catalase (CAT) and superoxide dismutase (SOD) activity (Singh, Singh, & Singh, 2015).
9. Proline content: Accumulation of proline in plants exposed to any stress (Hayat et al., 2012).
10. Conducting tissue (Gniazdowska & Bogatek, 2005).
11. Plant-water balance (Sheteawi & Tawfik, 2007),
12. Genetic material (Baziramakenga, Leroux, Simard, & Nadeau, 1997; Jensen et al., 2001).
13. Growth and improvement: Upon exposure to allelochemicals, the growth and development of plants are influenced. The immediately visible impacts include repressed or impeded germination rate; seeds obscured and swollen; reduced

extension of root or radicle and shoot or coleoptile; swelling or rotting of root tips; twisting of the root axis; discolouration, absence of root hairs; increase in number of seminal roots; diminished dry biomass accumulation; and reduced reproductive potential (Wu, Pratley, Lemerle, Haig, & Verbeek, 1998). These net morphological impacts might be auxiliary manifestations of essential processes, caused by an assortment of more specific influences acting at the cellular or less or more at the molecular level in the recipient plants (Duke et al., 2002). Impacts of allelochemicals on seed germination apparently intercede through an interruption of normal cellular processes instead of injury through organelles. Reserve mobilization, a mechanism that generally occurs quickly amid early phases of seed germination appears to be by all accounts deferred or diminished under allelopathic stress (Gniazdowska & Bogatek, 2005). The biological responses of recipient plants to allelochemicals show a threshold response and are known to be concentration (Gulzar et al., 2016; Ishak & Sahid, 2014). The responses are, typically, incitement or attraction at low concentrations of allelochemicals, and suppression or repellence at comparatively higher concentrations (Qasem & Foy, 2001). These prodigies have been broadly reported in allelochemicals from living plants, in allelopathic impacts from plant residues undergoing decomposition, and from the wide range of morphology to biochemical activity, comprising other growth-regulating chemicals and synthetic herbicides (Akemo et al., 2000). As recommended by Einhellig (2002), an essential impact of phenolic acids is on the plasma membrane, and this modification in the permeability barrier adds to various physiological alterations inhibiting growth and development.

References

Abhilasha, D., Quintana, N., Vivanco, J., & Joshi, J. (2008). Do allelopathic compounds in invasive Solidago canadensis sl restrain the native European flora? *Journal of Ecology, 96*(5), 993–1001.

Akemo, M. C., Regnier, E. E., & Bennett, M. A. (2000). Weed suppression in spring-sown rye (Secale cereale)–pea (Pisum sativum) cover crop mixes. *Weed Technology, 14*(3), 545–549.

An, M., Pratley, J., & Haig, T. (1998). Allelopathy: From concept to reality. In *Proceedings of the 9th Australian agronomy conference* (pp. 563–566). Wagga, Australia: Australian Agronomy Society.

Baar, J., Ozinga, W. A., Sweers, I. L., & Kuyper, T. W. (1994). Stimulatory and inhibitory effects of needle litter and grass extracts on the growth of some ectomycorrhizal fungi. *Soil Biology and Biochemistry, 26*(8), 1073–1079.

Batish, D. R., Singh, H. P., & Kaur, S. (2001). Crop allelopathy and its role in ecological agriculture. *Journal of Crop Production, 4*(2), 121–161.

Baziramakenga, R., Leroux, G. D., Simard, R. R., & Nadeau, P. (1997). Allelopathic effects of phenolic acids on nucleic acid and protein levels in soybean seedlings. *Canadian Journal of Botany, 75*(3), 445–450.

Baziramakenga, R., Simard, R. R., & Leroux, G. D. (1994). Effects of benzoic and cinnamic acid on growth, chlorophyll and mineral contents of soybean. *Journal of Chemical Ecology, 20,* 2821–2833.

Bertin, C., Weston, L. A., Huang, T., Jander, G., Owens, T., Meinwald, J., & Schroeder, F. C. (2007). Grass roots chemistry: Meta-tyrosine, an herbicidal nonprotein amino acid. *Proceedings of the National Academy of Sciences, 104*(43), 16964–16969.

Blum, U. (2002). *Soil solution concentrations of phenolic acids as influenced by evapotranspiration.* Abstracts of the Third World Congress on Allelopathy (pp. 56).

Bogatek, R., Oracz, K., & Gniazdowska, A. (2005). Ethylene and ABA production in germinating seeds during allelopathy stress. In *Fourth world congress in allelopathy.*

Celik, T. A., & Aslanturk, O. S. (2010). Evaluation of cytotoxicity and genotoxicity of Inula viscosa leaf extracts with Allium test. *Journal of Biomedicine and Biotechnology, 2010*(189252), 1–8.

Cheng, H. H. (1992). A conceptual framework for assessing allelochemicals in the soil environment. In *Allelopathy* (pp. 21–29). Dordrecht, The Netherlands: Springer.

Chou, C. H. (1980). Allelopathic researches in the subtropical vegetation in Taiwan. *Comparative Physiology and Ecology, 5*(4), 222–234.

Duke, S. O., Dayan, F. E., Rimando, A. M., Schrader, K. K., Aliotta, G., Oliva, A., & Romagni, J. G. (2002). Chemicals from nature for weed management. *Weed Science, 50*(2), 138–151.

Einhellig, F. A. (2002). The physiology of allelochemical action: Clues and views. In M. J. Reigosa Roger & N. Pedrol (Eds.), *Allelopathy from molecules to ecosystems* (pp. 1–23). Enfield, UK: Science Publishers.

Galindo, J. C., Hernández, A., Dayan, F. E., Tellez, M. R., Macias, F. A., Paul, R. N., & Duke, S. O. (1999). Dehydrozaluzanin C, a natural sesquiterpenolide, causes rapid plasma membrane leakage. *Phytochemistry, 52*(5), 805–813.

Gniazdowska, A., & Bogatek, R. (2005). Allelopathic interactions between plants. Multi site action of allelochemicals. *Acta Physiologiae Plantarum, 27*(3), 395–407.

Gulzar, A., Siddiqui, M. B., & Bi, S. (2016). Phenolic acid allelochemicals induced morphological, ultrastructural, and cytological modification on *Cassia sophera* L. and *Allium cepa* L. *Protoplasma, 253*(5), 1211–1221.

Hayat, S., Hayat, Q., Alyemeni, M. N., Wani, A. S., Pichtel, J., & Ahmad, A. (2012). Role of proline under changing environments: A review. *Plant Signaling and Behavior, 7*(11), 1456–1466.

Ishak, M. S., & Sahid, I. (2014). Allelopathic effects of the aqueous extract of the leaf and seed of Leucaena leucocephala on three selected weed species. *AIP Conference Proceedings, 1614*(1), 659–664.

Jensen, L. B., Courtois, B., Shen, L., Li, Z., Olofsdotter, M., & Mauleon, R. P. (2001). Locating genes controlling allelopathic effects against barnyard grass in upland rice. *Agronomy Journal, 93*(1), 21–26.

Khalid, S., Ahmad, T., & Shad, R. A. (2002). Use of Allelopathy in agriculture. *Asian Journal of Plant Sciences, 3*, 292–297.

Mohamadi, N., & Rajaie, P. (2009). Effects of aqueous eucalyptus (E. camadulensis Labill) extracts on seed germination, seedling growth and physiological responses of Phaseolus vulgaris and Sorghum bicolor. *Research Journal of Biological Sciences, 4*(12), 1292–1296.

Muscolo, A., Panuccio, M. R., & Sidari, M. (2001). The effect of phenols on respiratory enzymes in seed germination. *Plant Growth Regulation, 35*(1), 31–35.

Mushtaq, W., Ain, Q., & Siddiqui, M. B. (2018). Screening of alleopathic activity of the leaves of *Nicotiana plumbaginifolia* Viv. on some selected crops in Aligarh, Uttar Pradesh, India. *International Journal of Photochemistry and Photobiology, 2*(1), 1–4.

Mushtaq, W., Ain, Q., Siddiqui, M. B., & Hakeem, K. U. R. (2019). Cytotoxic allelochemicals induce ultrastructural modifications in *Cassia tora* L. and mitotic changes in *Allium cepa* L.: A weed versus weed allelopathy approach. *Protoplasma, 256*, 857. https://doi.org/10.1007/s00709-018-01343-1

Mushtaq, W., & Siddiqui, M. B. (2018). Allelopathy in Solanaceae plants. *Journal of Plant Protection Research, 58*(1), 1–7.

Pagare, S., Bhatia, M., Tripathi, N., Pagare, S., & Bansal, Y. K. (2015). Secondary metabolites of plants and their role: Overview. *Current Trends in Biotechnology and Pharmacy, 9*(3), 293–304.

Putnam, A. R. (1985). Weed allelopathy. *Weed Physiology, 1*, 131–155.

Qasem, J. R., & Foy, C. L. (2001). Weed allelopathy, its ecological impacts and future prospects: A review. *Journal of Crop Production, 4*(2), 43–119.

Reigosa, M. J., Souto, X. C., & Gonz, L. (1999). Effect of phenolic compounds on the germination of six weeds species. *Plant Growth Regulation, 28*(2), 83–88.

Rice, E. L. (1984). *Allelopathy* (2nd ed., p. 421). New York, NY: Academic Press.

Rizvi, S. J. (2012). *Allelopathy: Basic and applied aspects* (p. 23). Berlin, Germany: Springer.

Sheteawi, S. A., & Tawfik, K. M. (2007). Interaction. Effect of some biofertilizers and irrigation water regime on mung bean (*Vigna radiata*) growth and yield. *Journal of Applied Scence and Research, 3*(3), 251–262.

Singh, A., Singh, D., & Singh, N. B. (2015). Allelopathic activity of *Nicotiana plumbaginifolia* at various phenological stages on sunflower. *Allelopathy Journal, 36*(2), 315–325.

Singh, B., Jhaldiyal, V., & Kumar, M. (2009). Effects of aqueous leachates of multipurpose trees on test crops. *Estonian Journal of Ecology, 58*(1), 38–46.

Singh, B., Uniyal, A. K., & Todaria, N. P. (2008). Phytotoxic effects of three Ficus species on field crops. *Range Management and Agroforestry, 29*(2), 104–108.

Teerarak, M., Laosinwattana, C., & Charoenying, P. (2010). Evaluation of allelopathic, decomposition and cytogenetic activities of *Jasminum officinale* L. f. var. grandiflorum (L.) Kob. on bioassay plants. *Bioresource Technology, 101*(14), 5677–5684.

Weston, L. A. (2005). History and current trends in the use of allelopathy for weed management. *Cornell University Turfgrass Times, 13*, 529–534.

Wu, H., Pratley, J., Lemerle, D., Haig, T., & Verbeek, B. (1998). Wheat allelopathic potential against an herbicide-resistant biotype annual ryegrass. In *Proceedings of the Australian Agronomy Conference, Wagga, Australia* (pp. 567–571).

Chapter 8
Future Prospective

The allelopathic plants have been recommended as a suitable alternative for weed control under sustainable agriculture (Dahiya, Kumar, Khedwal, & Jakhar, 2017). Allelopathic plants smother weeds in the field following crop rotation (Dwivedi, Shrivastava, Singh, & Lakpale, 2012), cover or smother crops (Ch, Sturm, Varnholt, Walker, & Gerhards, 2016; Sturm, Peteinatos, & Gerhards, 2018), intercropping (Dhungana, Kim, Adhikari, Kim, & Shin, 2019), mulching (Abbas, Nadeem, Tanveer, Farooq, & Zohaib, 2016; Mabele & Ndong'a, 2019), and water extracts of allelopathic crop (Iqbal, Khaliq, & Cheema, 2019; Shahbaz, Sohail, Faisal, & Muhammad, 2018). The repressive effect of various allelochemicals in crops and trees is mainly ascribed to obstructed physiological and metabolic processes of a plant that has been used directly and indirectly for weed management (Cheng & Cheng, 2015; Farooq, Jabran, Cheema, Wahid, & Siddique, 2011). Allelochemical utilization for restricting weed growth is a realistic substitute for manufactured herbicides that do not have any harmful impacts (Bhadoria, 2011). Allelopathy is a new method offering numerous answers for the diminishing food accessibility under increasing worldwide population.

Regardless of the multifaceted research procedures, allelopathy as a science is quickly developing and the methodology exists as one segment of plant obstruction in nature. Additionally, its remarkable job in nature is currently well recognized. In spite of the fact that obstacles in plant–plant interference demonstrated to be a noteworthy hindrance in seeing how allelopathy works, investigation on allelopathy in recent years has expanded, however, diverse bunches of zones have still yet not been begun. There is an adequate scope to do research on donor and recipient plants. Various methodologies, like, phenotypic characters, biotechnology, physiology, life structures, plant source–sink relationship, supplement accessibility, inadequacy, biology, ecological factor, soil physical and synthetic contracts, and analysis of allelochemicals produced can be utilized for investigating and describing distinctive agrochemicals which are derived from characteristic sources. Considerable number of reports has been recorded about the effective utilization of allelopathic plants and their extracts into the yield fields as a substitute of synthetic herbicides to control weeds. Additionally, the consolidated use of reduced synthetic herbicides dose and allelopathic concentrates can provide control that is as powerful as that acquired from the standard portion of herbicides (Farooq et al., 2011). Hong, Xuan, Eiji, and Khanh (2004) worked with ten allelopathic higher plants (*Ageratum conyzoides, Bidens pilosa, Blechnum orientale, Eupatorium cannabinum, Euphorbia hirta, Galactia pendula, Leucaena glauca, Melia azedarach, Morus alba, and Tephrosia candida*) and unveiled that these species at 2 t·ha^{-1} resulted in a remarkable suppression of paddy weed growth and stimulated growth and production of rice. Among these species, *B. pilosa* and *T. candida* indicated the most noteworthy potential to diminish over 80% weeds and enhanced rice yields by up to 20%. From another study, Laosinwattana, Teerarak, and Charoenying (2012) reported that the powders of the *Aglaia odorata* leaves suppress the emergence and growth of *Digitaria adscendens, Trianthema portulacastrum*, and *Amaranthus gracilis* at a dosage of 1 t·ha^{-1}. A bioherbicide 'Porganic™' was then produced from the leaf extracts of this plant, which had a remarkable inhibitory effects on *Echinochloa crusgalli* and *Sphenoclea zeylanica* at a dosage of 10 kg·ha^{-1} (Laosinwattana, Huypao, Charoenying, Lertdetdecha, & Teerarak, 2013).

Xuan, Shinkichi, Khanh, and Chung (2005) reported that incorporation of 1–2 t·ha^{-1} of strong allelopathic plants could reduce 70–80% weed. The most significant obstruction for utilizing plant biomass is that water is required for its decomposition. Hence, it is very hard to apply plant biomass in areas where scarcity of water occurs. To surpass this problem, numerous researchers presently focus on allelopathic substances rather than the plant itself. The researchers expressed that such substances could reduce the load from plant residue supplementation. They stated that allelopathic substances with stronger activity on weeds could be used as a tool for new natural herbicides development. Recently, various compounds extricated from higher plants, for example, cineole, benzoxazinones, quinolinic corrosive, and leptospermones, have been economically applied in crop fields to control weeds (Barton, Dell, & Knight, 2010; Macías, Molinillo, Varela, & Galindo, 2007; Schulz, Marocco, Tabaglio, Macias, & Molinillo, 2013). In addition, these herbicides are not exactly enough to control an immense number of weeds, and also not viable to all weeds. In this way, looking for new common plant items viable for weed control is exceptionally significant. Isolation and recognition of new characteristic plant items may prompt the improvement of new regular herbicides. Utilization of allelopathic weed control through crop rotation, intercropping, cover cropping, mulches, residues, and water extract individually or in blend with other synthetic herbicides will provide sustainable crop production besides sustainable weed control because of productive effects of these procedures on soil fertility, organic matter contents, and ecosystem biodiversity. Moreover, endeavors to prompt industries to create allelochemical-based herbicides, investigating the allelopathy of unexplored fields, utilization of allelochemical hormesis, and conception of allelochemical mode of action will improve the adequacy of allelopathic weed control (Farooq, Abbas, Tanveer, & Jabran, 2019).

References

Abbas, T., Nadeem, M. A., Tanveer, A., Farooq, N., & Zohaib, A. (2016). Mulching with allelopathic crops to manage herbicide resistant littleseed canarygrass. *Herbologia, 16*(1), 31–39.

Barton, A. F., Dell, B., & Knight, A. R. (2010). Herbicidal activity of cineole derivatives. *Journal of Agricultural and Food Chemistry, 58*(18), 10147–10155.

Bhadoria, P. B. S. (2011). Allelopathy: A natural way towards weed management. *American Journal of Experimental Agriculture, 1*, 7–20.

Ch, K., Sturm, D. J., Varnholt, D., Walker, F., & Gerhards, R. (2016). Allelopathic effects and weed suppressive ability of cover crops. *Plant, Soil and Environment, 62*(2), 60–66.

Cheng, F., & Cheng, Z. (2015). Research progress on the use of plant allelopathy in agriculture and the physiological and ecological mechanisms of allelopathy. *Frontiers in plant science, 6*, 1020.

Dahiya, S., Kumar, S., Khedwal, R. S., & Jakhar, S. R. (2017). Allelopathy for sustainable weed management. *Journal of Pharmacognosy and Phytochemistry, 6*, 832–837.

Dhungana, S. K., Kim, I. D., Adhikari, B., Kim, J. H., & Shin, D. H. (2019). Reduced germination and seedling vigor of weeds with root extracts of maize and soybean, and the mechanism defined as allelopathic. *Journal of Crop Science and Biotechnology, 22*(1), 11–16.

Dwivedi, S. K., Shrivastava, G. K., Singh, A. P., & Lakpale, R. (2012). Weeds and crop productivity of maize + blackgram intercropping system in Chhattisgarh plains. *Indian Journal of Weed Science, 44*, 26–29.

Farooq, M., Jabran, K., Cheema, Z. A., Wahid, A., & Siddique, K. H. (2011). The role of allelopathy in agricultural pest management. *Pest Management Science, 67*(5), 493–506.

Farooq, N., Abbas, T., Tanveer, A., & Jabran, K. (2019). Allelopathy for weed management. In *Co-evolution of secondary metabolites* (pp. 1–6). Basel, Switzerland: Springer.

Hong, N. H., Xuan, T. D., Eiji, T., & Khanh, T. D. (2004). Paddy weed control by higher plants from Southeast Asia. *Crop Protection, 23*, 255–261.

Iqbal, N., Khaliq, A., & Cheema, Z. A. (2019). Weed control through allelopathic crop water extracts and S-metolachlor in cotton. *Information Processing in Agriculture.* https://doi.org/10.1016/j.inpa.2019.03.006

Laosinwattana, C., Huypao, J., Charoenying, P., Lertdetdecha, K., & Teerarak, M. (2013). Herbicidal activity of PORGANICTM, application and its potential used as natural post-emergence herbicide in paddy rice. In *Proceedings of the 24th Asian-Pacific Weed Science Society Conference* (pp. 376–382).

Laosinwattana, C., Teerarak, M., & Charoenying, P. (2012). Effects of *Aglaia odorata* granules on the seedling growth of major maize weeds and the influence of soil type on the granule residue's efficacy. *Weed Biology and Management, 12*, 117–122.

Mabele, A. S., & Ndong'a, M. F. (2019). Efficacy of guava (*Psidium guajava*) mulch allelopathy in controlling tomato (*Solanum lycopersicum*) weeds. *East African Journal of Agriculture and Biotechnology, 4*(1), 7–11.

Macías, F. A., Molinillo, J. M. G., Varela, R. M., & Galindo, J. C. G. (2007). Allelopathy—A natural alternative for weed control. *Pest Management Science, 63*, 327–348.

Schulz, M., Marocco, A., Tabaglio, V., Macias, F. A., & Molinillo, J. M. (2013). Benzoxazinoids in rye allelopathy-from discovery to application in sustainable weed control and organic farming. *Journal of Chemical Ecology, 39*(2), 154–174.

Shahbaz, K., Sohail, I., Faisal, M., & Muhammad, N. (2018). Combined application of sorghum and mulberry water extracts is effective and economical way for weed management in wheat. *Asian Journal of Agriculture and Biology, 6*(2), 221–227.

Sturm, D. J., Peteinatos, G., & Gerhards, R. (2018). Contribution of allelopathic effects to the overall weed suppression by different cover crops. *Weed Research, 58*(5), 331–337.

Xuan, T. D., Shinkichi, T., Khanh, T. D., & Chung, I. M. (2005). Biological control of weeds and plant pathogens in paddy rice by exploiting plant allelopathy: An overview. *Crop Protection, 24*, 197–206.

Printed in the United States
By Bookmasters